D0893463

COMMERCIAL
FISH FARMING

COMMERCIAL
FISH FARMING

WITH SPECIAL REFERENCE TO
FISH CULTURE IN ISRAEL

Balfour Hepher
Fish and Aquaculture Research Station, Dor, Israel

Yoel Pruginin
Former Chief Extension Officer—Aquaculture,
Ministry of Agriculture, Israel

A WILEY-INTERSCIENCE PUBLICATION
JOHN WILEY & SONS
New York • Chichester • Brisbane • Toronto

SH
151
.H53

Copyright © 1981 by John Wiley & Sons, Inc.

All rights reserved. Published simultaneously in Canada.

Reproduction or translation of any part of this work
beyond that permitted by Sections 107 or 108 of the
1976 United States Copyright Act without the permission
of the copyright owner is unlawful. Requests for
permission or further information should be addressed to
the Permissions Department, John Wiley & Sons, Inc.

Library of Congress Cataloging in Publication Data:

Hepher, Balfour, 1925–
    Commercial fish farming.

"A Wiley-Interscience publication."
Includes index.
    1. Fish-culture. 2. Fish-culture—Israel.
3. Carp. 4. Tilapia. 5. Gray mullets. I. Pruginin,
Yoel, joint author. II. Title.

SH151.H53   639.3'11   80-28593
ISBN 0-471-06264-2

Printed in the United States of America

10 9 8 7 6 5 4 3 2 1

EL
148069

PREFACE

Aquaculture is today considered an important source of production for meeting the world's increasing demand for protein. Aquaculture development projects are being initiated in many parts of the world, especially in the developing countries. In many cases, the main constraint to their success is a lack of knowledge of the basic principles and the technical skills involved.

Aquaculture in Israel, which has been developed over the past four decades, has had considerable success. Production has increased from an average of 1.5 tons/ha to almost 4.0 tons/ha, with the leading farms reaching over 7.0 tons/ha. This success has caused much interest and has led to many requests to share the knowledge gained in Israel. Unfortunately, the methods practiced have not been well documented, and most of the information is scattered in numerous papers in various publications and journals, both in Hebrew and in English.

Because of our recent involvement in teaching trainees from foreign countries who are faced with the same lack of documented information, we decided to solve this by writing a book that brings together most of the principles and practices, published and unpublished, on this subject. We soon realized that the general principles and techniques are applicable to many other regions besides Israel. Therefore, while the emphasis of this book remains on the Israeli experience, its scope has been broadened to include more general aspects. Some practices of aquaculture less familiar to us, such as catfish farming in the United States, or mariculture, have been omitted or are only briefly mentioned.

We would like to thank all our colleagues in Israel and abroad, as well as fish farmers, who commented on early drafts of this book and helped us, we hope, to make the final version a better one. Special thanks are due to the staff of the Department of Foreign Training of the Israeli Ministry of Agriculture for their help and encouragement. We would appreciate receiving comments or suggestions in order to help us improve future editions of this book.

BALFOUR HEPHER
YOEL PRUGININ

Dor, Israel
July 1981

v

CONTENTS

1 ———————————————— Introduction

1.1 WHY FISH CULTURE?

Early in the 1970s, the catch of fisheries in the oceans seems to have reached a peak and leveled off. According to the Food and Agricultural Organization of the United Nations (FAO, 1978), the total world fish catch in the oceans increased steadily from World War II until the beginning of the 1970s, but since then there has been no appreciable increase in this catch. World marine fish catch fluctuated between 56.8 and 64.1 million tons (Figure 1.1) from 1972 through 1977. If fish production in inland water is added to that, total world fish production in 1977 amounted to 73.5 million tons. The demand for fishery products, on the other hand, is constantly increasing. It is estimated (Pillay, 1973) that by the year 1985, this demand will reach about 107 million tons. This leaves a gap of about 35 million tons of fish. A major way to close this gap would be to increase fish production through aquaculture.

The annual production of fish by aquaculture is currently about 6 million tons, of which two-thirds are finfish (Pillay, 1979). This is only about 8% of world production but, contrary to the state of marine fisheries, it is constantly increasing. Further increases in fish production through aquaculture can be achieved in two ways: (1) by increasing the area of fishponds and the facilities of other aquacultural methods (enclosures, cages, raceways, etc.); (2) by increasing the production per unit area of existing aquaculture fisheries, mainly fishponds.

Pillay (1973) presented information on both pond area and fish production in 31 countries. The average yield for those countries was 1.5 ton/ha/yr. However, the most striking fact in the production figures is the large differences among the averages of those countries, which fluctuate from 55 kg/ha/yr to 6.6 ton/ha/yr. No doubt part of these differences relate to variations in climate, but a considerable portion can be attributed to management methods. Table 1.1 presents figures on yields that have been obtained in properly managed ponds. All except the last involved *polyculture* (the culture of different fish species

1

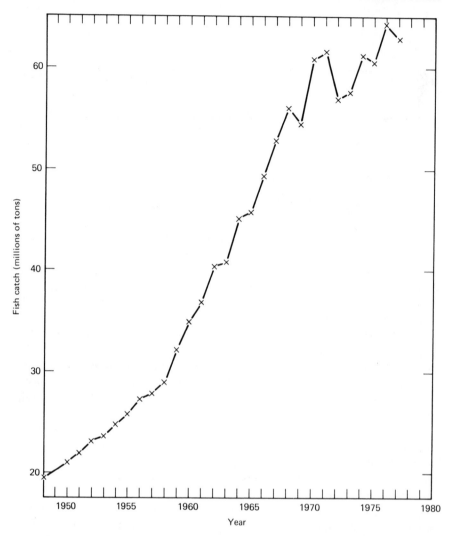

FIGURE 1.1. World marine fish catch for the period 1948–1977. (FAO, 1978.)

in the same pond). These figures demonstrate the potential of aquaculture, which is still to be developed.

Such high yields, however, can only be obtained through proper management based on knowledge of the biological processes in the pond and on the existence of a technical and organizational infrastructure such as regional water supplies and drainage, dependable supplies of fry, extension programs and training, and a marketing system.

TABLE 1.1 Fish Production in Properly Managed Ponds under Different Management Regimes

Management Regime	Country	Yield (ton/ha/yr)	Source
Fertilized ponds, no supplementary feeding	Israel	2.6–2.7	Yashouv, 1971
Heavily manured ponds, some feeding with rice bran	Taiwan	6.8–7.7	Tang, 1970
Heavily manured ponds, no supplementary feeding (extrapolated to 240 days)	Israel	4.6–7.8	Moav et al., 1977
Fish fed on protein-rich pellets	Israel	8.1–12.0	Moav et al., 1977
High stocking rates (over 14,000/ha), fish fed on protein-rich pellets (extrapolated to 240 days)	Israel	13.4–39.4	Sarig and Marek, 1974
Monoculture of carp in running water, Tanaka farm	Japan	1500–2000	Kawamoto, 1957; Brown, 1979; Tamura, 1961

1.2 COMMERCIAL FISH FARMING DEVELOPMENT

Two crucial considerations in planning aquaculture development are the size of the aquaculture unit and the form of ownership. An appreciable part of the fishponds existing in the world today, especially in the developing countries, have been constructed to supply protein for the diet of the farmer and his family. These "subsistence" ponds are usually small, not more than 100–500 m². Most attempts to develop such "subsistence" aquaculture or to increase the yield of the existing ponds have failed. The small scale does not justify the purchase of necessary equipment such as nets, and it does not provide any incentive for professional training. The yield of subsistence ponds is usually, therefore, very low, and thus the system conforms with the farmer's low standard of living and does not provide means for its change.

On the other hand, if the farmer sells his fish as a cash crop obtained from a larger farm unit, the result is usually an increase both in his standard of living and in fish production. Such a farm unit can be integrated with general agricultural activity, such as animal husbandry, utilizing the waste of the latter to increase production in the pond. It can also be integrated with the irrigation system of the farm, using fishponds as operational reservoirs. The fish farm can, of course, be a unit by itself unconnected to any other agricultural activity. In any case, the minimal size of such a farm should be such that income from it will justify the purchase of essential equipment and create an incentive for further training and professional advancement.

These principles, however, admit a wide range in the size of the farm and in the method of management, depending upon local economic structure and other local constraints such as: cost of investment; cost and availability of labor, water, and land; price of fish; and the extent of infrastructural support services. Where investment costs are high, land and labor are limited and costly, and the price of fish is high (e.g., Japan), the trend will be for intensification in order to achieve maximum yield per unit area. But where land, labor, and fish are inexpensive and feed is unavailable or costly (e.g., the Philippines), the trend should be for larger pond area, utilizing more of the natural food for fish production (see Chapter 7).

Another aspect affecting the management of the fish farm, and in many cases the fish yield that can be obtained from it, is the pond's physical characteristics and whether it can be drained or not. For maximum efficiency farm ponds must be drainable. However, in some regions of the world, especially in South Asia, aquaculture is based on existing water bodies rather than on ponds constructed especially for fish culture. These are usually village ponds of 0.3–1.0 ha, which cannot be drained by gravity. In most cases, therefore, these ponds are not drained at all. They are usually filled only by rain runoff during the monsoon. During the dry season water level drops, sometimes to only 30–40 cm, but rarely is desiccation complete.

Although high yields have been obtained from properly managed undrainable ponds, on the whole their fish yields are usually very low. The average yields of 0.6 million hectares of undrainable ponds in India was only 600 kg/ha (Aquaculture Development and Coordination Programme, 1976). This low yield seems to be due to these causes:

Uncontrolled Population of Pests.   The draining and refilling of a pond checks, at least temporarily, the development of the pest population of water bugs, tadpoles, frogs, and predacious fishes (see Sections 6.1.5 and 6.1.7). Nondrained ponds often maintain a permanent population of these pests, which may decimate the fish population stocked into the pond, especially when the fish are small.

Drastic Reduction in Water Volume During the Dry Season. Low water volume may cause an increase in the concentration of accumulated metabolites, especially when low water volume coincides with a high standing crop of fish. Low water volume can also cause an increase in water temperature beyond the optimal range for fish growth and thus a rapid decrease in growth rate.

Accumulation of Organic Matter in the Sediments. The draining of a pond results in the aeration of its bottom. This oxidizes reduced compounds that accumulate in the soil and increases the decomposition rate of the organic matter. In a nondrained pond this is not possible, and organic matter accumulates on the bottom. Due to the decomposition of organic matter oxygen is completely consumed and decomposition continues anaerobically, producing toxic compounds such as hydrogen sulfide (H_2S) and methane. These substances affect the productivity of the pond and reduce fish yields. The increase in organic content of the pond sometimes seems to be associated with blooms of blue-green algae which also hinder fish production.

Proper treatment and management of nondrained ponds can increase their yield considerably (see Section 7.3.4). However, the procedures involved are more suitable for smaller ponds and fish farms. In large commercial farms they create an extra burden and lower the efficiency of production. Therefore, where possible, drainable ponds should be constructed.

The infrastructural supporting services mentioned above—water supplies and drainage, fry supply, extension, and a marketing system—are essential for the development of commercial fish farming. All of them must be taken into account since the lack of any one may cause the failure of the aquaculture development program. Such a program, after setting national or regional production targets, may follow two alternative pathways:

1 Taking the existing structure of agriculture as a starting point and planning aquaculture accordingly. In many developing countries this means small farm units which will be entirely dependent on receiving the supporting services from outside. These services can be provided either by the government, by cooperatives, or by private commercial enterprises. Any development program for such small fish farm units must make a major effort to provide these infrastructure services.

2 By first defining the best ways to use available resources and the existing supporting services, and planning the farm units accordingly. These are usually larger units with various forms of ownership, such

as a cooperative, a governmental company, or a private enterprise. They usually involve the use of hired labor. It is obvious that supporting services, such as fry supply and marketing, are much better utilized by these larger units. Moreover, the larger they get, the more they are self-supporting with respect to these services. They can produce their own fry, employ people with sufficient professional qualifications, and even undertake some experimentation on the farm. If production is high enough, such large farms can also maintain an independent marketing system.

Technical operations on the farm may differ according to local socio-economic conditions. Every operation done by machines has its equivalent in manual labor. This is true for all aspects of pond operations, from the construction of ponds to fish harvesting. Management of the farm can thus be labor intensive or labor extensive, depending upon availability of labor and its cost. From a production point of view, it is more important, therefore, to understand the basic factors and principles affecting planning and management than to emphasize specific technologies. There is no doubt, however, that these technologies, when properly used, can facilitate some of the farm operations.

Commercial fish farming in Israel is in a unique situation which allows us to draw examples for some of the principles of fish culture and the technology involved. This is true both because of the climate and because of the adaptability of the aquaculture industry to changing economic conditions.

Israel has a subtropical climate with a clear distinction between a completely dry and warm summer and a cool, wet winter. Average daily water temperatures during the summer range from 18 to 30°C with daily fluctuations bringing the maximum daily temperature up to 32°C. These temperatures are comparable to those prevailing in tropical regions. The major difference is the length of the growing season. While fish growth in the tropics continues all or nearly year round, in Israel the growing season is only about 240 days (March 15 to November 15). Conclusions drawn from experiments and experience in Israel acquaculture can, therefore, be applied to a certain extent to tropical regions. When comparing yields, however, differences in the lengths of the growing seasons should be taken into account and production should be expressed on a daily basis (kg/ha/day).

In winter, daily average water temperatures in Israel range between 10 and 18°C, dropping occasionally during cold spells to about 5°C. This enables us to draw some conclusions which may be applied to colder regions, such as the effect of low temperature on the growth of carp, survival of tilapia, and other similar points.

Fish culture in Israel has developed from a monoculture system based on the common carp and adopted from European practices to polyculture systems based on methods similar, in principle, to those practiced in China. These methods have been adapted to conditions in Israel where intensification was motivated by the high cost of labor and limitations on water and land. It is clear that this development and the resulting achievements can serve as a model for commercial fish farming projects elsewhere. Methods and techniques cannot, of course, be transplanted and copied in regions with different environments and economic systems. However, the principles and considerations underlying the planning of farms, the planning of individual ponds, the choice of fish species, determination of the culture method, and so forth, can be adopted. These principles can be applied in a wide range of conditions.

1.3 DEVELOPMENT OF FISH CULTURE IN ISRAEL

Most fish farms in Israel are found on collective farms (*kibbutzim*) and are a part of more comprehensive agricultural activities. The planning of such a farm unit and its operation are, therefore, often affected by more general considerations of integration or competition with other agricultural activities on the farm. Land, water, and labor are the

FIGURE 1.2. A view of a fish farm on the Mediterranean coast of Israel.

main limiting factors in Israeli agriculture, and any agricultural activity, including fish culture, is evaluated by its efficiency in utilizing these resources, that is, profit per unit of land, unit of water, and man days of work. The decision whether to develop a fish farm or to use the same land and water for a field crop such as cotton depends on this evaluation. Naturally, other factors affecting profitability such as the cost of investment, the yield per unit of area and water, availability and costs of fish feed, demand and prices of market fish, and others are involved.

The development of fish culture in Israel is described graphically in Figures 1.3 and 1.4. Fish culture started with a privately owned experimental farm established in 1934 by immigrants from central Europe (Hornell, 1935). Only a few such privately owned farms were established and none has survived to the present. They could not withstand the vagaries of development, whereas fish culture in the kibbutzim thrived as parts of larger agricultural units.

The first kibbutz fish farm was established in 1938 at Nir David. The outbreak of World War II, with its increased demand for freshwater fish and the relative availability of land and water, enhanced the development of more fish farms until there were 30 by the end of the war in 1945 on a total area of about 800 ha. Since fish culture was brought from Europe, European methods were adopted at first: monoculture of carp in shallow ponds. Because the climate in Israel has a longer season for fish than that of Europe, yields were higher in Israel. Supplementary feed such as lupine seeds, various oil cakes, and cereal grains were used, but feed conversion ratios (weight of feed/weight gain of fish) were high, reaching 6 to 7 (Reich, 1952). After mastering the management techniques and adjusting stocking rates to the natural foods in the pond average yields increased from about 1.0 ton/ha in the first few years to about 1.5 ton/ha.

The water used in Israeli culture ponds was usually brackish and not suitable for crop irrigation. In addition the land on which the ponds were built was unsuitable for cultivation. The construction of new fishponds required, and still requires, permits granted only when the land and water have no alternative use.

During the mid-1960s, competition for water become intense, especially since the growing of cotton had been introduced. The amount of water supplied to each farm was and still is strictly rationed. Since cotton can also use marginal brackish water, the alternative value of water had to be evaluated, that is, the expected profit attainable by using a given amount of water for cotton was compared to that from fish farming. Subsequently, the amount of existing pond area leveled off, and even decreased, during the 1970s. A number of management

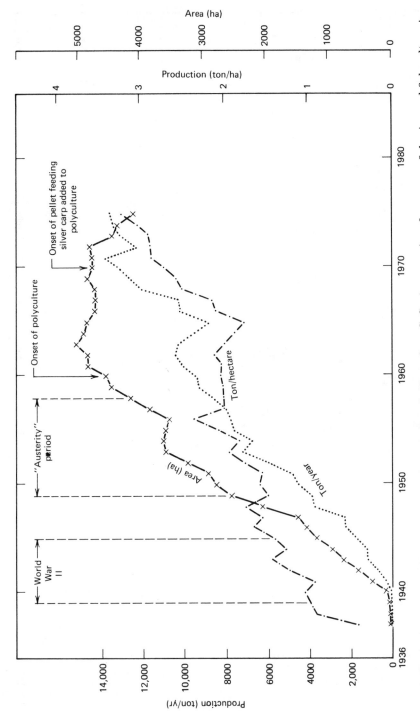

FIGURE 1.3. The development of aquaculture in Israel. (Data from S. Sarig—a series of papers on fisheries and fish culture in Israel in Bamidgeh.)

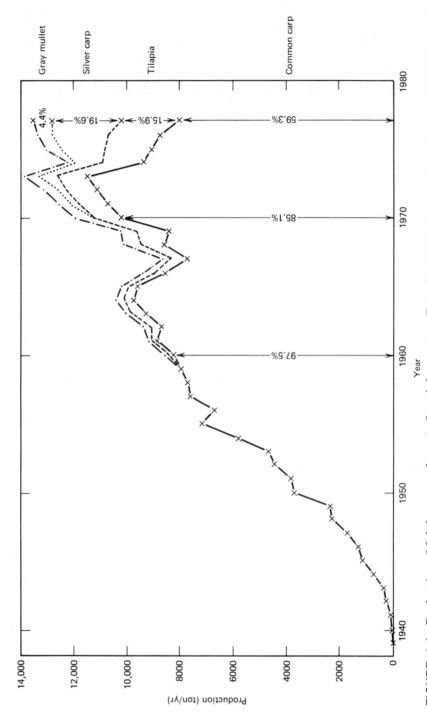

FIGURE 1.4 Production of fish by aquaculture in Israel, by species. (Data from S. Sarig—a series of papers on fisheries and fish culture in Israel in Bamidgeh.)

methods have been altered, however, resulting in an increase in yield per unit area, so that total production has increased, particularly during the last decade.

Efforts have been made to use new sources of water, such as runoff which occurs during the rainy winter months. For this purpose ponds were deepened from about 1 m to an average of 2–3 m. This seems to have contributed to an increase in yield (see Section 4.3). In the early 1960s polyculture was introduced by adding gray mullet (*Mugil cephalus*) and tilapia (*Sarotherodon aureus*). In the early 1970s silver carp (*Hypophthalmichthys molitrix*) and more recently white amur (*Ctenopharyngodon idella*) and bighead carp (*Aristichthys nobilis*) were added. As can be seen from Figure 1.4 these fish comprised an appreciable part of the yeild only in the 1970s, reducing the proportion of carp in the yield to 85.1% in 1970 and to 59.3% in 1977. This had a great effect on the yield.

The use of protein-rich pelleted fish feeds since 1970 has further increased the yield per unit area, since it enabled an increase in stocking rates, especially of carp and tilapia, which respond to supplementary feeding. A crucial factor was genetic improvement of the carp. This acted less dramatically than the other measures mentioned above, but no doubt it contributed to increased yield, especially by preventing inbreeding and introducing selected crossbred carp to the fish farms. As a result of these management methods the yield per hectare increased to 3.23 tons in 1977. Thus the fish culture system was adjusted to the changing demands and choices of consumers while high total production and profitability levels were maintained.

Labor was always in short supply in the kibbutzim, mainly as a result of the ideology of not employing hired labor and living on one's own work. This has recently become a limiting factor since intensification requires the handling of large amounts of fish. Technological devices for handling fish, such as elevators, feeders, and sorters, have been developed on the farms, resulting in a low labor input per unit of fish production although energy costs have increased.

As long as fish farming in Israel involved monoculture of common carp, or even the simple polyculture of carp and tilapia, each farm produced its own fry. The farm units were large enough to justify such activity. However, a number of developments have led to specialization in fry production:

1 Genetic improvement of common carp stocks required the breeding of selected genotypes for brood stocks, keeping them under controlled conditions to avoid contamination, and the production of crossbreeds from these selected lines.

2   The introduction of tilapia hybrids required maintenance of "pure" species for brood stocks used for hybridization. Since mixing of tilapia species is easy, strict prevention measures must be taken, usually involving maintenance of the brood stock indoors under completely controlled conditions.

3   The introduction of Chinese carps, which do not spawn spontaneously in ponds, required the employment of induced spawning techniques which demand special facilities and expertise.

This has led to a concentration of fry production on five farms which have established commercial hatcheries and specialize in breeding techniques. These farms supply fry to other fish farms as part of their overall fish culture activity.

It is obvious that current trends could not have developed without the contribution of research and extension services. Moreover, since the majority of farms depend for their fry supply on a relatively small number of hatcheries, it is essential to ensure permanent professional control over these hatcheries in order to: maintain the quality of fry produced; plan the annual production of fry carefully in advance so that the demands of the farms will be met and surpluses avoided or, in the case of a shortage, be distributed fairly among the farms; and control the price of fry.

The professional aspects of fish farming are controlled by research and extension personnel, while the organizational and commercial aspects are handled by the Fish Breeders' Association (FBA), a voluntary cooperative institution which includes all the fish farmers in the country. Fish are marketed through the cooperative marketing organization "Tnuva". However, the actual supply of fish to the various markets is organized and regulated by the FBA, which also controls the quality and size of marketed fish. This avoids a surplus in the markets and gives producers a fair share in marketing their produce. In this organized marketing system the gap between retail prices and the producers' return is quite small, ranging between 20 and 35%.

In view of its relatively highly developed state, fish culture in Israel has served as a major source for writing this book, though not the only one. Examples are also drawn from literature and experience in some developing countries.

2 ⎯⎯⎯ Selection of Site and Necessary Preconditions

2.1 SOIL

One of the advantages of fish culture over conventional agriculture crops is the ability of the farmer to utilize nonarable soils and water not suitable for irrigation. This is especially important where land and water are in short supply. Thus fishponds can be found on swampy soil as well as on sand dunes.

The most suitable soil for fishponds is a heavy clay soil where seepage is low. When ponds are constructed on sandy soil seepage is usually quite high, often reaching 10 cm/day and more. This decreases quickly due to blocking of the interstitial pores by organic matter produced in the pond and precipitated. In 1–2 years, the seepage drops to an acceptable rate of 1–2 cm/day. This process of blocking the soil pores can be accelerated by spreading about 10 m³ of cattle manure per hectare before filling the pond with water. This is repeated until seepage is reduced. In loamy soil the alteration of the soil structure through compaction by mechanical means, such as "sheep foot" or caterpillar tractor, may help in reducing seepage. Drying of pond bottoms allows oxidation and increases the rate of decomposition of organic matter sealing the soil pores. This in turn increases the rate of seepage. It is best in these ponds, therefore, to refill them with water immediately after draining.

Rocky and sandstone soils, as well as pebble beds, are not fit for pond construction. Fissures in the rocks and spaces between the pebbles and in the sandstone are too large to be blocked by organic matter. Small experimental ponds can be lined with polyethylene sheets or other materials, but this is too expensive for commercial ponds.

Areas with a high ground water level, such as tidal lands and marshes, may pose severe problems for pond construction in two ways: (1) usually these areas are too low to allow for complete drainage as required for proper fish farm management; (2) it is very difficult to use traditional earth moving equipment. Thus, if construction is not done manually, and this is possible only where manual labor is inexpensive, the construction of ponds requires special equipment and the costs may be prohibitively high.

Acidic soils are common in many parts of the world, for example, lateritic soils in the tropics, humic soils in the temperate zone, and acid sulfate soils in Southeast Asia. According to Singh (1980) acid sulfate soils cover over 15 million hectares in Southeast Asia alone. Acidity of these soils is caused by the presence of pyrite (FeS_2) in the soil. The necessary constituents for pyrite formation are sulfate, iron, metabolized organic matter, sulfer-reducing bacteria, and anaerobic conditions (Singh, 1980). These constituents are present in the coastal mangrove soils of Southeast Asia. The decomposition of the mangrove vegetation (especially their roots) takes up the available oxygen and creates anaerobic conditions. When the pyrite is exposed to oxygen it is oxidized to sulfuric acid, thus giving the soil a pH as low as 3.

Less common are soils of excessively high pH. These are formed mainly in swamps where soda ($NaCO_3$) is accumulated. Both low and high pH may be harmful for growth of fish and other aquatic organisms. Such soils, if their pH is not corrected, may limit yield or prevent fish culture entirely. According to Tang (1979) 100,000 ha of coastal fish farms in the Philippines are affected by acid soils and produce very low yields.

The effect of soil acidity on fish yields will depend to a large extent on the reaction of the water. If the pH of the water is basic, and there is a constant downward movement of water, leaching the acids, the effect on fish yields is small. In most cases, however, it is advisable to lime the soil either with agricultural lime [$Ca(OH)_2$] or limestone ($CaCO_3$). This technique should be examined for its economic feasibility according to the availability and costs of the lime and labor involved compared with the expected increase in fish yield. According to Swingle (1961), practical experience has shown that soils of pH 5 require approximately 2 tons of limestone ($CaCO_3$) per hectare and those with pH 4 require 4–6 ton/ha. Boyd (1979) developed a method for determining the lime requirements of pond bottoms from their pH and the change in pH in a buffered solution. Best results are obtained by spreading the lime over the entire pond bottom. The residual effect of liming depends on the rate of seepage through the bottom soil and the pH of the water. Usually liming must be repeated every year or two. Leaching of acid soils requires good drainage of the pond. Any site with acid (or alkaline) soils but no drainage is unsuitable for pond construction.

2.2 TOPOGRAPHY

Modern fish culture is impossible without control of the fish population in the pond. This demands a means of complete pond drainage. The most important consideration in selecting a site is the potential for pond drainage, preferably by gravity. This means that the bottom of the pond should be well above a receiving area such as a river or canal. Where this is not possible and the pond cannot be drained, special management methods must be employed (see section 7.3.4).

The best area for fish ponds is where the ground is level, or where there is a slight slope. The optimal slope is between 0.5 and 1%. In this case the pond can be perpendicular to the contour lines. The slope of the bottom of the pond will then follow the natural ground slope. This will reduce the amount of earth to be excavated, saving the need for excavating the deeper parts of the pond and making the planning of the pond much easier. Where the slope is steeper, the lower pond embankment must be high enough so that the shallow end of the pond will contain water to a sufficient depth for the fish. When the slope is very steep high and costly embankments must be constructed or narrow ponds must be formed, where the ratio between the length of the embankment and the water surface is high and construction is, therefore, expensive. It is not practical to construct embankments higher than about 3 m. This dictates that the maximum slope suitable for fishpond construction is about 2.5%. It is obvious that in a sloping area the pond axis should follow the contour lines.

The construction of ponds by throwing up earthen dams across natural hollows into which water is introduced from streams or runoff from the adjacent land is discussed in some publications on fish culture (Swingle, 1944; Woynarovich, 1975). Usually this is done when only one or a few ponds are constructed. These are very often made primarily for soil conservation and water storage for watering farm animals or for irrigation of crops. Fish can be cultured in such ponds but in most cases the stream bed is too narrow to contain the number of ponds required for a fish farm.

If any pond on the farm is to be constructed by damming a stream, it is preferable to select a stream with as little flood water as possible or to make the proper arrangements to divert excess water from the ponds. In any case a spillway is absolutely necessary to pass flood waters. This is usually built on one side of the dam, preferably on the undisturbed bank of the stream rather than on the dam itself.

2.3 WATER QUALITY AND TEMPERATURE

Poor water quality is detrimental to fish culture. Existing chemical constituents of the water and its temperature may determine not only which fish can be cultured but whether fish can be cultured at all.

Acidity and alkalinity of the water are important factors to be considered. Waters ranging in pH from 6.5 to 9.0 before daybreak are generally regarded as most suitable for pond fish production (Swingle, 1961; Alabaster and Lloyd, 1980). Excessively low or high pH beyond this range can cause lowered fish yields, and extreme pH values can be detrimental.

Water can be made acid or alkaline either naturally, by percolating through acid or alkaline soils, or by a direct discharge of wastes. In the second case the acidity or alkalinity also may be associated with toxic compounds such as certain organic acids or ammonia. These may modify the effect of the pH. We shall deal here only with naturally occurring pH.

Waters of low pH are common in humid regions, where water percolates through soils low in calcium and rich in humic acids. Such waters can be found in large parts of Africa, Southeast Asia, and Europe. Water with pH below 4.5–5.0 is toxic to most warm-water fishes (Swingle, 1961; Alabaster and Lloyd, 1980). Variability in the effect of pH on fish at this range depends on species, size of individual fish, and environmental conditions such as concentration of carbon dioxide or the presence of heavy metals such as iron.

Acid water with a pH range of 5.0–5.5 can be harmful to fish reproduction since eggs and fry are more sensitive to low pH than are larger fish. At a pH range of 5.5–6.5 fish production is low, either because of a direct effect on the fish or because of an effect on fish food organisms. From Alabaster and Lloyd (1980) it appears that the relation between pH per se and growth of fish is unclear, but there is a general agreement that low pH affects the productivity of the pond. Neess (1949) stated that acidity impedes the recirculation of nutrients by reducing the rate of decomposition of organic matter and inhibiting nitrogen fixation. Pruginin (FAO/UN, 1965) reported that ponds with pH below 6.5 in Uganda gave a very poor production. Common carp grew only few grams in 6 months. Similar results have been reported by Swingle (1961) for bluegill (*Lepomis macrochirus*) in Alabama.

Water of excessive alkalinity may also be harmful to fish. One should distinguish here, however, between high pH resulting from photosynthesis and that caused by environmental factors. The uptake of carbon dioxide during the photosynthetic process results in an increased pH. This is especially prominent in productive small bodies of water such as ponds rich in phytoplankton. In such ponds there is, usually, a strong daily fluctuation in pH. At daybreak pH is lowest, and it is highest in the late afternoon when photosynthesis reaches its peak. pH values of 9.5 are quite common in productive ponds, and pH can even reach 10–10.2 in ponds receiving sewage effluents. No harmful effects on warm-water fish have been observed in these conditions.

Swingle (1961) recommends, therefore, that pH measurements to determine the suitability of such waters for fish culture be made before daybreak. Better information can be obtained if a full daily pH cycle from sunrise to sunset is recorded.

Eggs and young fry may be affected by pH higher than 9.0 (Elster and Mann, 1950), but most authorities agree that fish are affected only when the pH reaches 10–10.5, and that pH 11 is lethal to fish (Swingle, 1961; Alabaster and Lloyd, 1980).

Water of low pH (5.5–6.5) may be corrected by liming to neutralize the acidity. The more acid the water, the more lime is required to neutralize it. The amount of lime required for neutralizing the water depends not only on the pH, but also on the chemical composition of the water, especially the concentration of calcium bicarbonate [$Ca(HCO_3)_2$] and its relationship with carbon dioxide and carbonates (for a more thorough discussion of this system see Hutchinson, 1957). Being a salt of a strong base and weak acid, calcium bicarbonate serves as a buffer which prevents sudden changes in pH through the addition of strong acids or bases such as lime.

The concentration of bicarbonates and carbonates in water is expressed as alkalinity. This is measured by the amount of acid required to change the pH of the water to 4, as indicated by the change of color of methyl orange indicator. This alkalinity is usually reported as ppm calcium carbonate ($CaCO_3$) equivalent to the acid used to titrate the water. Another way to express alkalinity is to report the volume of acid of a given strength that was used for titration. In Germany this was called *Saltzäurebindungsvermögen* (SBV) (Schäperclaus, 1961).

Low pH is usually associated with low alkalinity and thus low buffer capacity. Schäperclaus (1962) and Hickling (1962) stated that there is a relationship between alkalinity and pond productivity. The lower the alkalinity the lower the productivity and vice versa. It is difficult, however, to differentiate the effect of alkalinity from that of pH.

The amount of lime required to neutralize pond water also depends on the form of lime applied. When quicklime (calcium oxide, CaO) is used, much smaller amounts are required than when slacked lime or agricultural lime [calcium hydroxide, $Ca(OH)_2$] or limestone (calcium carbonate, $CaCO_3$) are used. To have the same effect the proportion that must be used is 1:1½:2, respectively. In view of the complexity of the system, it is quite understandable why the amounts of lime the literature gives as required to neutralize the water of fish ponds vary so much. According to Schäperclaus (1961) ponds of low alkalinity in central Europe require 250 kg/ha CaO or 500 kg/ha $CaCO_3$, applied to the water, often in addition to liming the pond bottom with 250–500 kg/ha CaO (or 500–1000 kg/ha $CaCO_3$). Mortimer (1961) recommends the use of 1675 kg/ha of lime in African fish ponds. This amount is to be spread on the water surface at monthly intervals. Marr et al (1966) state that in small ponds in Africa where intensive manuring is being done, lime should be spread over the surface once a month at a rate of 170–220 kg/ha. The best way to determine the amount of lime required for a specific pond or farm is to titrate the pond water to neutrality and calculate the equivalent amount of lime to be added. Liming is usually a repeated operation and, therefore, it adds to the production costs. This must be taken into account when deciding on the pond site. Analysis of pH should be among the first tests to be conducted when a site is being considered.

Water turbidity is caused by a high content of suspended solids such as clay minerals. Turbidity also can be caused by excessive production of phytoplankton (algal bloom). This, however, will be discussed in Section 11.1.2. Turbidity can affect fish directly by killing them, reducing their growth rate, or preventing their reproduction. Turbidity can also affect fish indirectly by reducing the natural food available to the fish.

The major mechanical effect of suspended solids is through the injury of the gill structure. The harmful effect will depend on the species of the fish and their resistance, and on the nature of the suspended solid particles, their hardness, and their angularity (Alabaster and Loyd, 1980). Wallen (1951, quoted from Alabaster and Lloyd, 1980) kept several species of fish in water containing clay. Most individuals of all species endured maximum turbidities of 100,000 mg/l for a week or more. According to Nikolsky (1963) the most pronounced mechanical effect of suspended particles occurs when the water contains about 4% by volume of solids.

A not lesser damage to fish is caused by decreasing the abundance of natural food in the pond and consequently fish productivity. Suspended solids decrease light penetration into pond water and thus primary production. Since the food web in the pond depends to a large extent on the organic matter produced through photosynthesis, reduced primary production results eventually in reduced secondary production. Shehadeh (1975) states that high water turbidity in African ponds, brought about by land runoff during the wet season, reduces primary production and oxygen levels in these ponds. Turbidity can also be directly harmful to zooplankton due to clogging of their filter-feeding apparatus and digestive organs. Alabaster and Lloyd (1980) quote a number of authors to show the harm caused to Cladocera and Copepoda by clay minerals at concentrations of 100–500 mg/l.

When a pond contains an excessive concentration of suspended solids, it is likely that some of the factors mentioned above will affect its productivity, although it is difficult to determine the relative importance of each in every case. Van Someren and Whitehead (1959) have found an apparent lowered growth of *Tilapia nigra* in turbid ponds in Africa as compared to clear-water ponds. So did Buck (1956, quoted from Alabaster and Lloyd, 1980) in an analysis of the production of fish in 39 farm ponds in the United States. These ponds were stocked with largemouth black bass (*Micropterus salmoides*), bluegill (*Lepomis macrochirus*), and red-ear sunfish (*Lepomis microlophus*). After two growing seasons the yields of fish were:

Clear ponds (25 mg/l suspended solids)	161.5 kg/ha
Intermediate (25–100 mg/l suspended solids)	94.0 kg/ha
Muddy ponds (100 mg/l suspended solids)	29.3 kg/ha

High turbidity can be reduced by the use of settling ponds. Shehadeh (1975) states that this can alleviate the problem. Jhingran (1975) recommends scattering gypsum over the entire water surface, at a rate of about 200 kg per 1000 m³ of pond water. He states that sometimes it is necessary to repeat the application of gypsum with an additional 50 kg per 1000 m³, about 6 weeks after the first one. It is obvious that if the gypsum is to be applied repeatedly, it will affect the operation costs, and the economics of its use should be evaluated.

Many species of fish used in culture tolerate wide ranges of salinity. This enables the culturist to utilize brackish or even sea water if the proper species are selected. Common carp, which is the main fish cultured in Europe and Israel, can tolerate salinities up to about 5.0‰ without any appreciable effect on growth. The lethal concentration of salts for carp is about 11.5‰ (7000 ppm Cl^-) (Soller et al., 1965). Some tilapia species can adapt to much higher salinities, even up to sea water. Grey mullet (*Mugil cephalus*), a marine fish, can tolerate fresh water with no difficulty, and so can other species such as the milkfish (*Chanos chanos*). It is apparent that while some adaptation of fish species to salinity may be necessary, salinity does not limit fish culture.

Due to evaporation salt concentration in ponds increases constantly. Where evaporation is high (5–7 mm/day), and there is no addition of water to make up for the loss, salt concentration can double within 150–200 days (Yaalon, 1964). This should be taken into consideration when brackish water is used, since salinity may increase above the tolerance limit of some species. In such cases either the species composition should be chosen to conform with increasing salinity or the water should be frequently changed.

Wastewater, after proper treatment, is sometimes used in fishponds (see Section 9.5). The organic matter contained in this water is an important nutrient resource which can increase the production of natural food in the pond and, consequently, increase fish yields. There are a number of factors that should be considered when using such water in fishponds. Sanitary considerations are of prime importance, but not less important are public acceptance of the fish, the effect of wastewater on the oxygen regime in the pond, and the effect of detergents and other pollutants. In any case care should be taken not to introduce toxicants that may kill the fish. Industrial wastes should be avoided and, if that is not possible, the flow of such toxicants should be controlled at their source. It is not advisable to introduce wastewater directly into the pond. It should first pass through a control tank or pond. An alternative source of fresh water should always be available, either for use when the wastewater is too hazardous, or to hold fish after harvesting for a depuration period.

Water temperature may also be a limiting factor in fish production, and it is important in determining the species of fish that can be cultured. Where water is cold (below 10°C) in winter, special arrangements should be made to maintain tilapia over the winter. In some regions the water is too cold for this group of species even in the summer (see Section 5.3).

Water temperature can determine the management method even for fish, such as common carp, that can tolerate temperatures. The growth rate is usually much slower at lower temperatures, and the culture period required to bring the fish to market size is therefore longer. Due to the short growth season in central and eastern Europe, it takes 2–3 years to culture carp to a market size of 1–1.5 kg. Backiel and Stegman (1968) found a significant correlation between the annual common carp yield in a Polish fish farm and the number of days with a temperature above 20°C during July and August. For every day in which the temperature reached above 20°C during those months, the production of common carp on natural food alone increased by 4–4.7 kg/ha, depending on rearing conditions. The number of warm days with temperatures over 20°C during summer in Europe is rather limited. In the Polish farm mentioned above there were only 34 to 77 such days during April to September. In southern Europe the number of days with favorable temperatures may be larger but the growth season is still limited. Cool weather and a short growth season also can be found in the tropics at higher altitudes.

Water temperature can be specific to a certain source. Cold-water sources in warm climates can permit the culture of cold-water fish. On the other hand, a warm-water spring in a colder climate allows a longer growth season and higher annual production. Carp and tilapia spawning in such warm waters can be obtained earlier than on the surrounding farms. Production of early fry can be of commercial value if there is a demand for early hatched and nursed fry.

2.4 WATER QUANTITY

The minimum yearly requirement of water per hectare of fish pond includes the initial amount required to fill the pond and that required to compensate for losses by seepage and evaporation. For a fishpond of an average depth of 1.5 m, the initial amount of water required is 15,000 m³/ha. In Israel the average loss by seepage and evaporation during the growing season (240 days) is estimated to be roughly 1–2 cm/day. If the water level of the pond is allowed to drop by the end of the growing season to a depth of 1 m, about 5000 m³/ha are saved, and the amount of water required to top off the pond is about 20,000–45,000 m³/ha/yr. Thus, assuming an average pond depth of 1.5m, the total minimum quantity of water required for a hectare of fishponds for filling and topping is between 35,000 and 60,000 m³/ha/yr. The total amount of water available will thus determine the area of the planned fish farm.

Loss of water by evaporation and seepage will vary from one area to another. According to Huet (1970), the average loss due to evaporation in Europe is about 0.4–0.8 cm/day, while in tropical regions though variations are high, the rate can reach 2.5 cm/day. Stanhill (1963) estimated the water loss due to evaporation from fish ponds in different regions of Israel as 164.5–174.8 cm/yr, or an average daily evaporation of 0.45–0.48 cm/day. There may, of course, be large differences in evaporation in different seasons of the year. Parsons (1949) measured the loss of water, both by seepage and evaporation, for a pond at Auburn, Alabama. He reported the seepage losses to be 243 cm/yr or an average of 0.66 cm/day and evaporation losses as 118 cm/yr or an average of 0.32 cm/day. Total average losses in this case reached about 1 cm/day. Of course the addition of water by rain should also be taken into account.

When the water supply is seasonal, such as from a rainy period surface runoff, this fact must be considered when planning the depth and shape of individual ponds, as well as of the entire farm. Arrangements must be made in the catchment area to intercept the runoff and direct it into the ponds either by gravity or by pumping. The volume of water that can be accumulated will determine the size of the fish farm.

If it is desired to drain the pond completely after each growing period (about 100–140 days), as is the case with most farms in Israel, arrangements should be made to recycle the water. The water can be transferred to neighboring ponds or drained into a common reservoir from which it can be pumped and redistributed among the ponds as required.

The distance from the water source to the ponds is important from an economic point of view: the longer the pipes or canals, the higher the investment costs.

3 Planning the Fish Farm

3.1 SIZE OF THE FARM UNIT

When no limitations are imposed by the amount of water and land area, the main factors that determine the size of the farm unit are marketing possibilities and conditions, manpower, and equipment. The economic calculation which takes into account marketing potential and prices on the one hand, and the investments for construction of ponds, water supply, and aeration, as well as alternative usage of water and land on the other hand, will determine the most suitable method of management and its degree of intensification, and thus the size of the farm unit (see Chapter 7).

Certain operations on a fish farm, such as sample weighings, thinning out, transport of fish, etc., require teamwork. The smallest team capable of handling these operations would consist of three workers. These workers can handle a farm of 30–50 ha, depending on the number of ponds. During the growing season additional workers are required for harvesting, sorting, and packing the fish, while in winter one worker is usually sufficient.

The equipment required for carrying out the operations mentioned above, which serves the entire farm unit, consists of a set of nets of different sizes and mesh, an elevator for loading fish, one tank for transporting fish on the farm, one tank for transporting fish to market, and, if separate feeders are not used, a feed blower (see Chapter 8). This equipment can serve a farm of up to 50 ha producing about 250 tons of fish.

When more intensive culture methods are applied and the same level of production can be attained from a smaller area, the minimum size of the farm unit is reduced. In polyculture of common carp with tilapia, silver carp, and mullet, the extra work involved in sorting and packing fish should be taken into account. A larger number of workers is required for harvesting, sorting, and packing, and some additional equipment is required. These considerations do not hold where labor is inexpensive and does not play an important role in overall expenses.

The size of individual ponds determines the overall ratio between the total area of the farm and the area covered by water. The smaller the ponds, the greater is the proportional area occupied by embankments and drainage ditches. With the common size of rearing pond in Israel (4–5 ha), the embankments and drainage ditches constitute about 20–25% of total farm area.

3.2 DIVISION OF THE FARM AREA

The farm consists of rearing ponds and auxiliary ponds, which include ponds for segregation of brood stock, spawning ponds, fry nursing ponds, fry holding ponds, and storage ponds for marketable fish. These auxiliary ponds are much smaller than the rearing ponds—they usually range from 0.1 to 1 ha in area—and may serve different functions in different seasons. The same pond may be used for carp spawning in spring, fry nursing in summer, storage of market-size fish in autumn, and as a fry holding pond in the winter. Since the amount of fish harvested from the rearing pond does not necessarily correspond to the market demand for fish, storage ponds in which live fish are kept before marketing are necessary in all commercial fish farms.

The area of the storage ponds is determined by the amount of fish expected to be stored. A pond can hold about 1 kg/m^3 with no special arrangements, and up to 15 kg/m^3 if a continuous water supply or aeration is provided. The ponds should be easily accessible by vehicles in order to facilitate loading. Where tilapia are stored for marketing during winter at temperatures between 13 and 17°C, the storage ponds should be small enough to accommodate only one normal market consignment. In larger ponds where only a portion of the fish are marketed, handling of the remaining tilapia in the pond may cause injuries and deaths. In all, the auxiliary ponds may constitute about 15% of the total pond area of the farm.

3.3 WATER SUPPLY AND DRAINAGE

Each pond requires its own separate water supply. Where water is in short supply, it is advisable to have special arrangements to enable the transfer of part, if not all, of the water contained in one pond to an adjacent pond by gravity during draining so as to reuse and save the water. However, this cannot be the only arrangement for water supply. The kind of water supply system, whether pipes or open canals, should be decided upon after analyzing local economic realities. Open canals are always less expensive than pipes. In open canals, however, it is more difficult to prevent wild fish from entering the pond, and the canals require more work for maintenance. The diameter of inflow pipes will be affected by four main factors: (1) the amount of water available, (2) rate of flow, (3) the time required for filling the pond, and (4) the price of the pipeline system. The most common pumping units used in fish farms in Israel have acapacity of 200–300 m^3/hr. The pipe system required to distribute this amount of water to the ponds usually has a diameter of 15–25 cm (6–10 in.).

Where water is in short supply and its reuse is essential, this should be ensured by proper integration between the drainage and supply systems. An operational reservoir for pumping of drained water into the supply system should be constructed.

4 _____ Pond Construction

4.1 SIZE OF REARING POND

The most important factors affecting and determining rearing pond size are:

Cost of Construction.   The larger the pond, the smaller will be the construction cost per unit area. Grizzel (1967) gives the following example for the effect of size on costs of catfish ponds in the United States. The embankments of a 16 ha pond (40 acres) will occupy about 2 ha, while the embankments of the same area divided into four ponds averaging 4 ha each will occupy 3 ha, and the cost per hectare will be about double.

Time Required for Filling and Draining Pond.   The larger the pond, the longer it takes, at a given water flow, to fill or drain. During the favorable growth season this may mean a significant loss of yield. For instance, if a pond of 4 ha is harvested during the summer, at a draining rate of 1000 m³/hr, it will take about 2 days to drain the last 50 cm (the draining is done only during the day when it can be attended). At that time fish are not fed and normally do not grow. At a water flow rate of 300 m³/hr (see Section 3.3), the time required to fill the pond to a depth of 30–50 cm, which is the minimum safe depth for stocking, will be 67 hr or about 3 days. The total time lost will thus be about 5 days. Since the yield in productive ponds may be as high as 30–50 kg/ha/day, this may mean a loss of 250 kg/ha of fish per harvesting. At water flow rates usually found on fish farms in Israel, it is not advisable therefore to construct a pond larger than 8–10 ha.

Expected Yield According to Planned Management Method.   The higher the yields per unit area, with the resultant need to handle larger amounts of fish, the more important becomes the yield factor in considerations of pond size. Due to the sensitivity of some of the fish in the pond to harsh conditions in the harvesting sump (see Section 4.4.1) and the danger of possible fish loss during harvesting, it is desirable to complete harvesting as quickly as possible. It should take no more than 1 day per pond. The maximum amount of fish which can be handled by a normal size team and equipment in one day is 10–20 tons. This determines, to a large extent, the size of the pond. Using the usual management methods practiced in Israel, when protein-rich feed is used, the standing crop of fish which can be expected at harvesting is 2–4 ton/ha. This means that a harvest of 10–20 tons can be obtained from a 4–5 ha pond, which should, therefore, be considered the optimal size. With increasing intensification of culture and the consequent increase in yields per unit area, and with increases in the number of species cultured in polyculture systems, the tendency is to reduce the size of ponds to 2–3 ha. When even more intensive methods are employed, yields may reach 20 ton/ha/yr, and standard harvests may reach 10 ton/ha. The optimum size of the pond is then even smaller, reaching 1 ha or even less.

The considerations presented above affect the upper limit of pond size, but smaller ponds are sometimes constructed in spite of increased costs. Woynarovich (1975) states that in Nepal, since land suitable for fishpond construction is expensive, economical pond size is only 0.5–1.0 ha.

4.2 SHAPE OF REARING PONDS

The main factors affecting the desired shape of rearing ponds are:

1 The ratio between the length of embankments and the area covered by water; this affects the cost of construction.

2 The topography of the area.

3 The anticipated method of fish harvesting.

In a square pond where the ratio of water area to the length of embankment is highest, the cost of construction is lowest. When there are no topographical limitations and fish harvesting is to be carried out by draining, or if the ponds are small, square ponds are recommended.

The effect of slope of the area on the shape of ponds has been discussed previously (see Section 2.2). It is obvious that the greater the slope, the stronger will be its effect on the shape of the pond. The pond must then be narrower and, therefore, more expensive.

The method of harvesting will also affect, and sometimes determine, pond shape. If harvesting or intermediate thinning out of the fish population is to be done by seining, pond width should not be greater than 100 m. Greater widths demand expensive changes in the length and strength of nets, increased manpower, and towing equipment. This may negate the cost saving in a wider pond.

When demand or automatic feeders (see Section 8.2.2) are installed in the pond, fish will crowd by the feeders, and it is then possible to harvest them by netting in only a small area by the feeders. This, in turn, necessitates the positioning of feeders at the shallower end of the pond to enable netting. If the pond is not harvested by seining across its banks, as in the last case, or if the pond is harvested by draining, then clearly there is no importance attached to pond shape, which can be determined by cost and topography alone.

4.3 POND DEPTH

The minimum pond depth in Israel is usually 80 cm at the shallow end. When a pond is shallower, aquatic plants, such as common reed (*Phragmites communis*) and cattail (*Typha* sp.) develop. These plants decrease the effective area of the pond (see Section 8.5) and impede seining. Otherwise, a depth of 80 cm is sufficient for rearing fish. In order to maintain this water level throughout the growing season a sufficient and permanent water supply is required.

When the water supply is seasonal, it is necessary to increase pond volumes by increasing depth so as to accumulate enough water during the wet period to suffice for the entire growing season. The average depth can then reach about 2.5–3.5 m.

Some fish farmers claim that increased pond depth also has a beneficial effect on the growth, rate of fish, and though this has not been proved experimentally, it is in agreement with Chinese practice. Tang (1979) provides information on the grouping of Chinese fishponds according to depth and level of fish production (Table 4.1).

Increase in fish yield in deeper ponds may be due to a better temperature regime in the ponds. Because of their greater surface/volume ratio, peaks in temperature during the day are not as high as in the shallow ponds, where the temperature can reach about 32°C in the early afternoon hours of a summer's day. Daily temperature fluctuations are also smaller in deeper ponds. This may be important in areas having a warm climate since water temperature may exceed the optimum level for fish growth. In colder areas, such as Europe, the problem is how to attain higher water temperatures more rapidly. A shallow pond is, therefore, recommended (Huet, 1970).

An additional explanation for the beneficial effect of deep ponds on fish growth may be the dilution of metabolites excreted by the fish.

TABLE 4.1 Grouping of Chinese Carp Fishponds Based on Water Depth and Level of Fish Production[a]

Water Depth (m)	Level of Fish Production (kg/ha/yr)	Remarks
1.2–2.0	< 7500	Most ponds were this depth before the start of the fish farm land consolidation program. There are still extensive areas with ponds of this depth.
2.0–3.0	~ 7500	Most ponds have been constructed at this depth since the fish farm lands were consolidated.
3.0–4.0	> 7500	Most high-production ponds belong in this depth category.

[a] From Tang (1979).

These may have an adverse effect on fish growth. It seems, however, that there is a maximum depth beyond which the growth of the fish decreases, though this has not yet been determined.

The embankment should be about 0.5 m higher than the water level. When a dam is constructed across a stream, the top of the dam should be at least 1 m above pond water level if the area occasionally receives even moderate amounts of flood water. If large amounts of flood water must be handled, the top of the dam should be about 1.5 m above water level (Swingle, 1944). A suitable spillway should be provided in either case.

Deep ponds require some technological changes. For example, nets have to be deeper. If a pond is deeper than 2 m, seining becomes difficult, and the pond should have a shallow area to permit seining for sample weighings and thinning out fish when the pond is full. The elevator should also be longer in order to reach the deeper level (see Section 8.4.3).

4.4 POND STRUCTURE

4.4.1 The Pond Bottom

In order to drain water and concentrate fish in the deeper part of the pond, a minimal bottom slope of 0.1–0.2% is sufficient. For complete drainage, the deepest part of the pond near the outlet (the *monk*) should be deeper than the bottom level by as much as possible (up to a maximum depth of 1–1.5 m). This deepest part of the pond (the *harvesting sump*) should not be larger than 0.1 ha, and in small ponds it can be much smaller. The most important characteristics of the harvesting sump are its depth and volume. It should be able to contain all the fish in the pond just prior to harvesting. The position of the harvesting sump is, therefore, very important. It should be protected from siltation and should have a supply of fresh water and/or aeration for the period that the fish are concentrated in it. Some farmers protect the harvesting sump from silting in by building low levees made of sandstone or gravel around it. The water and fish pass through openings in these levees. In some parts of the world concrete bottoms and even side walls are used in sump areas.

4.4.2 Water Inlet

The water inlet is usually located near the outlet and the harvesting sump, so as to supply water when the fish are concentrated in the sump after the pond has been drained for harvesting. Locating the inlet at the deepest part of the pond also enables restocking shortly after starting refill (when the water in this area has reached a depth of 30–50 cm). A number of growing days can be gained in this way in large ponds where filling may require an extended periof of time. When the inlet is located at the shallow end of the pond immediate restocking is dangerous because the fish tend to swim against the current and become exposed to predators, such as birds and animals, at the shallow part of the pond. For the same reason, and also in order to avoid embankment erosion, the water inlet piple should extend into the pond beyond the foot of the embankment. The area where water falls into the pond is sometimes paved with stones to prevent the water from creating a depression in the bottom, and thus a pool when the pond is drained completely.

If the incoming water is not especially poor in oxygen there is no need to make special arrangements for spreading the water, such as waterfalls, sprinklers, etc. In some cases, as with cold water rich in oxygen, this may even reduce the oxygen content of the water (see Section 11.1).

When wild fish are abundant in the inflowing water, their intrusion into the pond should be prevented since they may be harmful (see Section 8.1). For this purpose permanent filters are sometimes installed in the water system, especially when the water is supplied through open channels (for less durable filters see Section 8.1). Filters are best made of gravel chambers constructed on the water inlet channel. Woynarovich (1975) gives details on the construction of such chambers. They are built as a gravel barrier through which the water passes. Since the structures reduce water flow, filters must be two to three times wider than the channel itself in order to pass the desired quantity of water. To maintain water flow through the filter, the gravel must be periodically removed and washed. A "reverse" gravel filter through which water passes from the bottom upward is very efficient if care is taken to clean it frequently. The frequency of cleaning depends on water turbidity.

4.4.3 Drainage and Water Outlet

A basic condition for well-managed fish culture is the total drainability of the ponds. To ensure this, the level of the bottom of the outlet pipe should be at least 30 cm above the drainage ditch to which it leads. The drainage system comprises a *monk*—the structure that regulates the water level in the pond—or any other type of water outlet, and a pipeline passing underneath the embankment and leading into the drainage ditch. Drainage is also essential for fish culture in a pond constructed by damming a river.

The diameter of the drainage pipeline should be sufficient to enable draining in not more than 1–3 days. In ponds of 4–10 ha, the diameter of the pipeline is usually 30–35 cm (12–14 in.). In smaller ponds 20–25 cm (8–10 in.) will be adequate. The pipeline is usually made of asbestos-cement or concrete. Those made of asbestos-cement are more costly, but they are also more durable and do not crack easily at the joints, as do the shorter segments of concrete pipe.

According to Grizzell (1967), most catfish farmers in the United States prefer a turn-down drainpipe for their ponds. Clearly this is practical for small ponds where a relatively small diameter pipe of up to 15 cm (6 in.) is sufficient for draining the pond. According to the same author, it takes about 2.5 days to drain a pond containing 6167 m^3 (5 acre-feet) of water with a 180 cm (6 ft) head, using a pipe of such a diameter (15 cm). This is much too long if the danger to the fish, both from predators and anoxia, during these days is considered. It seems therefore that larger diameter drainpipes should be installed, and then the monk is preferable. If a drainpipe is used, a screen should always cover the end of the drainpipe inside the pond to prevent loss of fish when water is discharged.

The most common water outlet is the monk (Figure 4.1), which is a square structure about 65 × 65 cm built in the pond at the foot of the embankment slope and attached to the drainage pipeline. Three of the monk's walls are built of concrete or bricks. The fourth is an open side blocked by wooden sluice boards inserted into slots provided for this

FIGURE 4.1. A concrete monk and drainage pipe.

purpose in the inner sides of the two opposite side walls. An additional pair of slots, parallel to the first one, serves for inserting a screen to prevent the escape of the fish while draining the pond. The second pair of slots can also serve for inserting an additional row of wooden boards. The space between these two rows of boards can be filled with clay or sawdust to stop any water leakage. This is done mainly in deep ponds where water pressure near the bottom is relatively high. Deep ponds are usually drained only once a year.

All the monks on the farm should be of a uniform width so that the wooden boards and screens are interchangeable. The sluice boards of the monk are 25 cm thick and 15–20 cm high. The edges of the boards are notched into "half-lap" to fit each other better and to prevent the passage of water between them. Sometimes two bolts are fitted to each board to serve as "handles" in order to ease removal of the boards without breakage. If only one row of sluice boards is used the slots are sealed by packing strips of jute burlap or other sealing materials into them after inserting the boards. This is done with the aid of a blunt chisel.

A screen is fitted on top of the sluice boards. This screen is also used when the pond is drained. The best screen is a grillwork made of round iron bars. A screen made of holes or a meshed wire tends to clog easily and is hard to clean. Each farm should have screens of different widths between the bars. 1 cm for nursing ponds and 2–3 cm for rearing ponds with large fish.

Some variations of the monk have been recently introduced. New materials such as plastic are being used both for the walls and sluice boards. One variation used instead of the concrete monk is a steel pipe of large diameter (35–40 cm = 14–16 in.) cut in half lengthwise. U-shaped slots are welded to the inner sides of the edges for the wooden boards and screens.

The monk is attached to a concrete base plate that is at least 1.0 × 1.0 × 0.2 m thick, to prevent it from sinking into the sediment. The base plate is level with the bottom of the drainage pipeline. When casting the plate, a slot for the lowest sluice board should be made for improved sealing.

The best location for the monk is near the leeward embankment, relative to the prevailing wind. Wind induces currents in the upper layers of the pond water, causing undercurrents in the opposite direction. The silt which is brought up from the bottom and the banks by wave action is thus carried away by the undercurrent and deposited along the opposite windward bank which is protected from the wind. If the harvesting sump is positioned along the windward embankment it will soon be silted up.

Locating the feeding site at the harvesting sump also helps in preventing silting up of the sump. Constant burrowing by the fish prevents silt from settling in the sump.

If the difference in elevation between the bottom of the sump and the drainage channel is too small and restricts the depth of the sump, it is advisable to reduce the slope of the bottom of the pond or even level it entirely so as to obtain increased sump depth.

4.4.4 Embankments

The width of the embankments' crest depends, to a large extent, on the types of vehicles and equipment that must pass over it. The main embankments serving the passage of heavy vehicles and equipment such as supply trucks, fish tanks, etc. should be wide enough (usually about 6 m) to allow for safe traffic. Secondary embankments serve for the passage of only smaller vehicles such as tractors for hauling the seine net. These embankments can therefore be narrower—about 4 m wide. Embankments that are not used for passage of vehicles at all can be still narrower, but with modern heavy construction equipment it is hard to build a well-packed embankment narrower than about 3.5 m. The height of the embankment will depend, of course, on pond depth but should be calculated to be about 0.5 m higher than anticipated water level.

The inner slope of the embankment has to withstand erosion caused by wave action and should be more gradual than the outside slope. The usual slope is of a height/width raio of 1 : 3. A steeper slope with a 1 : 2 ratio can be constructed if it is fortified by stones. This can be done with large stones or any soft stone such as sandstone or soft basalt. The outer slope of the embankment can be steeper and reach a height/width ratio of 1 : 1.5. It is obvious that an embankment dividing two ponds has two "inner" slopes and both should have a height/width ratio of 1 : 3.

The best equipment for constructing embankments is heavy earth-moving machinery which spreads the earth in thin layers and compacts these layers during construction. Scrapers seem to be most suitable, while bulldozers which pile the earth rather than spreading it do not compact it enough. An embankment constructed by a bulldozer is apt to become eroded much more quickly by wave action. In order to reduce erosion of embankments and prolong the depreciation time of the pond, the embankments should be protected not only when they are steep, as discussed above, but also when they have a slope of 1 : 3. This can be done with stones, sandstone, or soft basalt or by planting grass. It is most important to protect the leeward embankment as it erodes first. Covering the inner slope with a strip of stones 1 m wide, half of which is submerged when the pond is full, has been found to be adequate (Pruginin and Ben-Ari, 1959). Thickness of the stone layer should be 20–30 cm. Grass should be of a variety that spreads quickly, such as Kikuyu grass or Bermuda grass. Planting should be done immediately after completing construction or less favorable weeds will dominate.

4.4.5 The Process of Pond Construction

The first stage in the construction process involves general planning. For this purpose a topographical map of scale 1:2000 to 1:2500 is required. General features of the farm such as division of the area into various ponds and their types (rearing or auxiliary ponds), the number of ponds, size of the ponds, the orientation of the ponds, the width of the embankments, and the drainage and water supply systems (Figure 4.2) are all worked out on this map.

The next stage involves detailed planning of each pond. For this purpose a topographic work plan of larger scale is required. This will show where and how much to excavate and where the earth should be moved. For this purpose a survey has to be made. The pond area is

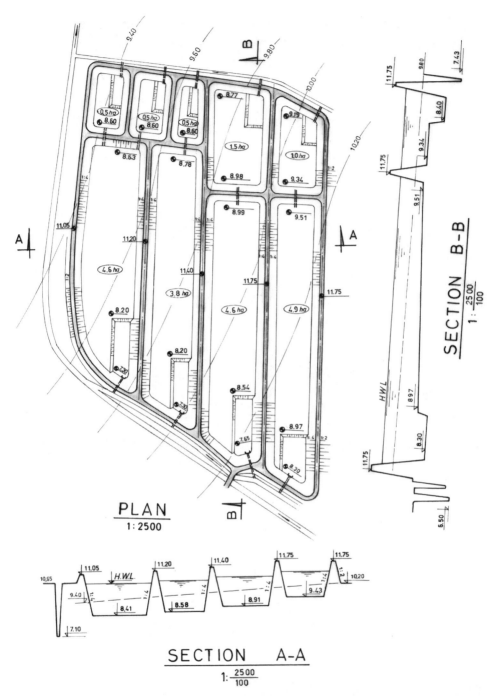

FIGURE 4.2. A plan of a fish farm area before construction.

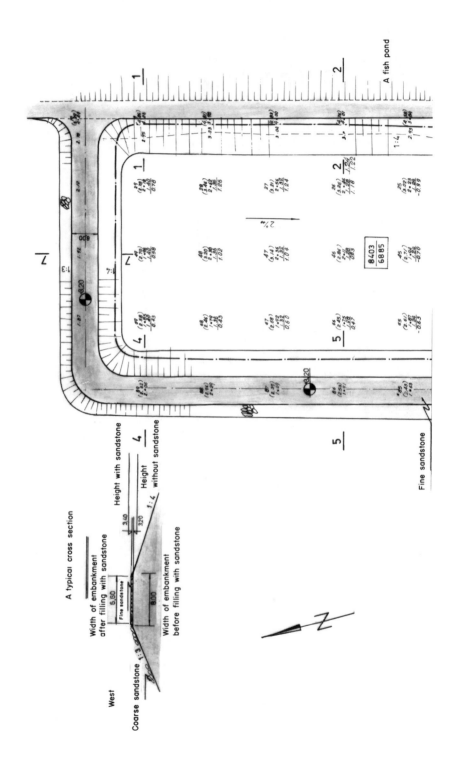

A typical cross section

Width of embankment after filling with sandstone

Height with sandstone

Height without sandstone

Width of embankment before filling with sandstone

West

Coarse sandstone

Fine sandstone

A fish pond

Fine sandstone

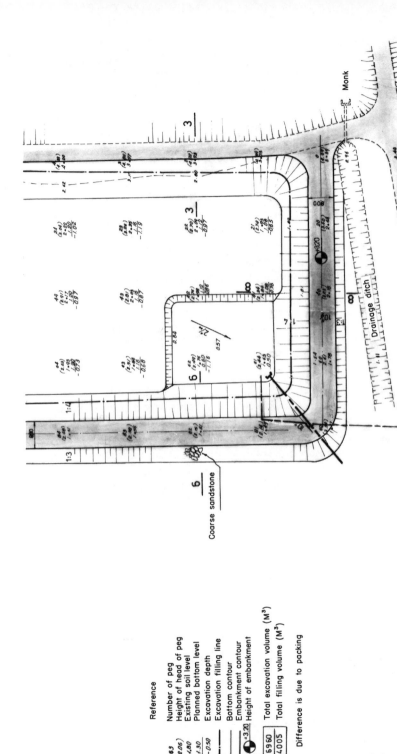

FIGURE 4.3. A working plan of a fishpond.

Reference

63	Number of peg
(2.06)	Height of head of peg
1.80	Existing soil level
1.30	Planned bottom level
−0.50	Excavation depth

Excavation filling line

Bottom contour

Embankment contour

+3.20 Height of embankment

6960	Total excavation volume (M³)
4005	Total filling volume (M³)

Difference is due to packing

Coarse sandstone

Monk

Drainage ditch

divided into 20 × 20 m squares by pegging the ground at these intervals. The peg line should be parallel to the direction of the longest embankment of the pond. If the ground surface between any two pegs is not regular and changes drastically by reason of a ditch or mound, a peg is added at the place of change. These pegs form a network of coordinates in which each peg is marked by a letter and a number. The ground level at each of the pegs and the levels of the peg heads are measured, and the elevations are noted on the work plan of the pond.

Next, the amount of earth required to construct the embankments is calculated from the general plan. This is done by determining the amount of earth required for 1 m of embankment (this is equal to the volume of a 1 m trapezoidal cross-section of the embankment having the planned dimensions) and multiplying by the total length of embankments. The packing coefficient of the soil should be taken into account. This coefficient depends on the type of soil and the kind of construction equipment used. The heavier the equipment, the greater is the packing coefficient, usually 20–50%. This means that when 1 m^3 of earth is required for the packed embankment, 1.2–1.5 m^3 should be provided.

Generally, about one-third to one-quarter of total pond depth is obtained by excavation. The rest is obtained by dumping earth on the embankments and building them up. The amount of earth which can be excavated is calculated and compared to the amount required (again, taking into account the packing coefficient). If these amounts do not balance, tentative excavation depth is changed accordingly and the amount of earth obtained is recalculated. According to Pruginin and Ben-Ari (1959) 2500–4000 m^3 of earth are required per hectare for construction of a 4 ha pond.

After excavation depth is calculated, the details are transferred to the working plan (Figure 4.3). The depth of the planned bottom from the head of the pegs is marked. The amount of earth dug out from each square between pegs is calculated, and the place on the embankment where the earth should be deposited is designated. This should be as near as possible to the place of excavation. The contours of the embankment's base are pegged in the field and construction is started according to the working plan.

At the onset of construction vegetation and roots should be removed. These should not be put into the embankment since organic material will rot and leave behind hollows which become weak points in the embankments and are the first to erode. Either the topsoil over where the embankment should be is removed to a depth of 5 cm or, instead, this area can be well plowed in order to tie the embankment to the base soil and prevent seepage.

The drainage pipeline is laid before starting construction of the embankment. The location and length of the pipeline are designed on the working plan. The pipeline should be covered by earth which is then carefully and methodically compacted by hand tools to prevent leakage along the line. Such leakage may lead to bursting of the embankment in the area of the pipeline. An additional precaution is the construction of an antiseep collar made of concrete or metal in the embankment around the pipeline and perpendicular to it, extending to both sides. The pipeline should be covered with enough earth to protect it against damage from the heavy machinery passing over it.

During the work, special attention should be given to the shape of the embankment to make sure that its width at all heights corresponds to the plan. If the embankment is constructed too narrow, this cannot be corrected later.

5 ——————— Main Fish Species ——————— Cultured in Ponds

The number of fish species presently cultured in ponds is considerable. Hora and Pillay (1962) list about 70 species of fish cultured in the Indo-Pacific region alone. This list does not include species which are cultured in other regions and those which have been introduced to fish culture since then. We can not discuss each of these species, and so we concentrate here on groups that have had, or have recently gained, wide distribution and seem to be adaptable to the different environmental conditions found in various parts of the world. Other species, some of which constitute the main fish cultured in certain regions [such as the milkfish (*Chanos chanos*) in the Philippines and Indonesia, or the major carps in India], are omitted.

It must be emphasized here that successful performance of a fish species under one set of conditions, whether natural or in ponds, does not necessarily ensure its success under a different set of conditions. For example, *Sarotherodon vulcani* in Lake Rudolph, Africa, reach a size of up to 7 kg, but they do not show any advantage over other *Sarotherodon* species when introduced into fishponds. Big-mouth buffalo (*Ictiobus cyprinella*) were introduced into Israel from the United States in 1956. The fish grew very slowly and were highly susceptible to infestation of *Lerneae cyprinacea*. A similar fate occurred in tench (*Tinca tinca*) introduced to Israel from Europe in the 1940s. They grew very slowly compared to common carp, did not reach market size within 1 year, and were susceptible to *L. cyprinacea* and, therefore, failed. Much was expected from the introduction of the pejerrey (*Basilichthys bonariensis*) to Israel from Argentina in 1959. While this fish does breed spontaneously and naturally in ponds and in the Kishon Reservoir into which it was introduced, it has an extremely slow growth rate, is very susceptible to *Lerneae*, and has, therefore, also failed.

5.1 COMMON CARP *(Cyprinus carpio)*

One of the most important fish cultured in the world today is the common carp. According to FAO (1978) worldwide production of carp in 1977 amounted to 310,000 tons, more than any other freshwater species. This figure does not relate to production through aquaculture alone, but there is no doubt that fish culture comprises a major portion of the total.

The common carp seems to have originated in Central Asia and spread to both east China and northwest Europe (Balon, 1974). It appears to have been domesticated independently in China some millenia ago and in Europe probably during Roman times. As a result there are a number of carp varieties known in the world. The Chinese common carp, which has spread all over the Far East, is sometimes known as the "big belly" carp (Hora and Pillay, 1962). It is a scaled, hardy fish which can withstand harsh conditions. It breeds early—at about 6 months of age—and the gonads develop to relatively large proportions (hence the name "big belly"). The domesticated European varieties are usually more "round," having a larger depth/length ratio. There are four different varieties with respect to scale cover ranges, including: (1) completely scaled; (2) partially scaled "mirror carp" (which is also known in some countries as the "Israeli carp") (Figure 5.1); (3) one row of big scales on each of its lateral sides; (4) virtually scaleless—the "leather carp." The last two varieties have a much lower growth rate (see Section 6.1.1) than the two mentioned first.

The common carp is an omnivore that feeds on various foods of plant and animal origin. When feeding on natural foods produced in the pond, young carp consume mainly protozoa and zooplankton such as copepods, cladocerans, and other large zooplankton. Soon, however, when the common carp is about 10 cm in length, it begins feeding on bottom fauna. It burrows in the mud and sucks in insect larvae, worms, molluscs, etc. A preferred important food organism is chironomid larvae.

When supplied with supplementary feed, the carp digests carbohydrates well and accepts a wide array of feedstuffs having a relatively low protein content such as cereals, legumes, as well as proteinous feeds such as various oil-cake meals, slaughterhouse residues, minced trash fish, etc. Not all of these feedstuffs have the same nutritional value, and their effect on the fish growth and feed conversion ratios may vary widely.

Common carp spawn readily in captivity when maintained in ponds. The spawning season varies with climate. Carp spawn throughout the year in tropical climates, with a peak from January to April. In

FIGURE 5.1. Mirror carp, *Cyprinus carpio*.

colder regions spawning becomes much more seasonal. In subtropical climates carp spawn from April to June, and in temperate regions they spawn from May to June. The "big belly" carp becomes mature at about 6 months old and 25 cm in length in warm climates, and somewhat later in more moderate climates. The European varieties mature at a later age—about 1 year in warm climates (e.g., Israel) and in their third or fourth year in colder climates (e.g., Germany and northeast Europe). Fish size does not constitute a limitation on spawning, at least in warm climates, and carp can spawn even when they weigh only 200 g if they have reached sufficient age. In Israel this may cause problems with "wild" spawning in rearing ponds stocked in spring with fry from the previous spring. For this reason carp are also spawned later in the summer. This may take place in late July or August, which is not the natural spawning season of carp, and spawning therefore sometimes has to be induced by pituitary injections (see Section 6.1).

5.2 CHINESE CARP

This group of carp, most of which are herbivores, is native to China, and until the middle of this century it did not spread much beyond that country because the carp do not breed in captivity and the only source of fry was the large rivers of China. Since induced breeding of these carp has been achieved, they have spread to many parts of the world and take an increasingly important part in polyculture systems. Only three of these carp will be dealt with here.

5.2.1 Silver Carp (*Hypophthalmichthys molitrix*)

Though a freshwater river fish, silver carp (see Figure 5.2) are able to live in slightly brackish water and are suitable for cultivation in fishponds. They do not breed spontaneously in ponds, however, as do common carp. Up until the 1950s, silver carp culture was confined to

FIGURE 5.2. Silver carp, *Hypophthalmichthys molitrix*.

China and Southeast Asia where fry were caught in large rivers such as the Yang-tze and its tributaries and the West River.

Silver carp feed mainly on phytoplankton as small as 30–40 μm by filtering these microscopic algae through their gill filter apparatus. According to Woynarovich (1975) a fish of 250 g body weight can strain 32 l of water per day through its gills. It is quite common for a fish of 500–600 g to grow 10 g/day and more.

The ability of silver carp to feed on small algae makes this fish especially suitable for polyculture. It feeds on a trophic niche usually not utilized by other fish such as the common carp (bottom feeder), mullet (detritus feeder), and others. Moreover, through its feeding habits it improves the environmental conditions in the pond by controlling blooming of the algae (see Section 7.3.1).

Escessive algae may be a hazard since any mass die-off that occurs in the pond will enrich the water with organic matter which increases oxygen demand by bacteria upon its decomposition. Silver carp usually prevent the build up of such excessive algae concentrations.

Silver carp can eat only finely ground supplementary feed. It does not, therefore, take particulate feed offered common carp and tilapia. It can, however, consume leftover feed which has disintegrated into fine particles and which may not be taken by other fishes. Thus silver carp further decrease the load of organic matter and the oxygen demand associated with it.

Due to the large number of intermuscular bones and to some the insipid taste of its flesh, the silver carp is not appreciated in some markets. The demand and, correspondingly, price for it are low. Another drawback for the fish farmer is the habit of the silver carp to leap over nets when seined. It is difficult, therefore, to sample or harvest it during the season. It is more easily harvested after draining the pond. The advantages gained by culturing this fish, however, make up for these difficulties.

The inability of silver carp, as well as other Chinese carps, to spawn spontaneously in captivity can be considered an advantage since these fish do not spawn "wildly" in the rearing ponds. On the other hand, special hatchery installations and know-how are required for induced spawning (see Section 6.3), and these are not always available.

5.2.2 Grass Carp *(Ctenopharyngodon idella)*

Like all Chinese carp, grass carp (see Figure 5.3) do not breed spontaneously in ponds and are reproduced outside China only by induced spawning.

The grass carp is a very fast growing fish when given proper feed, including grass. The fish feeds on various kinds of weeds, both soft and hard. The amount of grass eaten depends to a large entext on the water temperature. Above 20–25°C it can eat large amounts of grass—over 50% of its own body weight per day. The grass carp readily eats supplemental feed such as grains and pellets. However, it seems that the grass carp's utilization of these feeds is lower than that of the common carp. Porath et al. (1979) indicate that when grass carp are fed on pellets containing 25% protein their growth rate is much below par. A much higher growth rate is attained when the fish are fed on duckweed (*Lemna minor*). On this feed a fish of 400 g grew 4.3 g/day, as compared to only 2.4 g on pellets. Comparison of the feed conversion rates also shows the advantage of the duckweed. On the basis of wet weight of feed, it was 37.3 for the duckweed and 3.6 for pellets, but if the ratio between the dry weight of the feed to the wet weight of fish gain is taken, that of duckweed was only 1.2 as compared to 3.2 for pellets. A more pronounced difference was found in PER (protein

FIGURE 5.3. Grass carp, *Ctenopharyngodon idella.*

efficiency ratio, which is the weight of protein consumed per weight of fish gained). With duckweed it was 0.38 as compared to 0.9 with pellets.

5.2.3 Bighead Carp (Aristichthys nobilis)

The bighead carp is also a very fast growing fish. When introduced to Israel it grew to a weight of 1.5 kg during 5 months (June to November) in its first year. This fish, closely related to the silver carp in its feeding habits, is also a filter feeder. However, it filters larger organisms—mostly zooplankton such as rotifers, copepods, and cladocerans and large algae such as *Aphanizomenon* or *Microcystis* colonies. It is more tame than the silver carp, does not leap as high as the latter when the pond is seined, and is thus much easier to catch.

5.3 TILAPIA

The tilapias are all members of the family Cichlidae. They all have oblong body shapes with long dorsal fins which have 23–31 fin spines and rays. The nose has one nostril on each side.

The genus *Tilapia* included two subgenera which differed mainly in their way of reproduction. The fish of the one former subgenus have smaller, more numerous eggs which adhere, after spawning, to a substrate. The parents fan and guard the eggs and embryos until hatched. In the other former subgenus, one or both of the parents collect the eggs into their mouths after spawning for incubation and brooding. Some years ago Trewavas et al. (1972) and Trewavas (1973) gave reasons for dividing these two subgenera into separate genera. The former retains the generic name *Tilapia,* while the latter has been assigned the name *Sarotherodon.* Since this book is intended for practical purposes, and since most of the literature on the tilapias still carries the "old" generic name, the common name of the whole group is referred to here as "tilapia."

The tilapia species have become increasingly important in fish culture, especially in warmer climates. According to FAO (1978), total world production of tilapia (both *Tilapia* and *Sarotherodon* species, but excluding other Cichlids) reached 197,000 tons in 1977. Here again only part of this production was obtained through aquaculture, but that portion is constantly increasing.

The number of tilapia species cultured in ponds, both experimentally and on a commercial scale, is quite large. Huet (1970) mentions

16 species. Only three, however, have gained wide distribution: *Sarotherodon mossambicus, S. niloticus,* and *S. aureus.* These are discussed below in greater detail.

Tilapias are tropical warm-water fishes, most of which originated in Africa. The optimum temperature range for their development is 25–30°C. All are sensitive to low temperatures, with the lethal limit being 9–13°C, depending on species. This susceptibility of tilapia to low temperature limits their culture to places where winter water temperatures do not drop below the above-mentioned values, or to places where measures can be taken to prevent exposure to lethal temperatures. This can be done by using a warm-water supply or covering tilapia holding ponds with a greenhouse. Most of the tilapias are euryhaline and can live in brackish waters, and some even in sea water (Kirk, 1972).

The main problem in the culture of tilapias is their proliferation. They breed easily at an early age (3–6 months) even when still small, and they have multiple spawns during the year. This can increase the fish population in ponds to such an extent that stunting occurs. To overcome this problem it is necessary either to use species that grow fast and reach market size before they breed, or to rear monosex populations (see Section 7.3.3).

In most of the tilapias, the males have a much greater growth capacity than the females, even when reared separately (Hickling, 1968; Pruginin, 1968a; Shell, 1968). Mabaye (1971) and Fryer and Iles (1972) confirm this fact in a number of species and attribute this characteristic to genetic causes. However, it can also be associated with the spawning of the females. The females continue to spawn at frequent intervals, even if the eggs are not fertilized. Thus, energy is diverted from growth to egg production. In a mixed population, when the eggs are fertilized and develop, the females do not feed during the mouth incubation and brooding period, which is a considerable drain on body reserves (Chimits, 1955; Huet, 1970). A monosex male population is therefore preferable. In practice this can be obtained in three ways: (1) sexing the fish by differences in their genital papilae (see Section 7.3.3), (2) sex reversal at an early age by feeding hormones (see Section 6.2.4), and (3) crossing two species of *Sarotherodon* to obtain only male hybrids or a high percentage of males—90% and over (see Section 6.2.2). Not all crosses result in the desired high percentage of males. Some result in a lower proportion of males. The males produced by the hybridization are fertile and can breed with any female *Sarotherodon* present in the pond. Thus special precautions are needed to keep the brood stock uncontaminated.

There is a controversy about heterosis for improved growth rate in tilapia hybrids. A number of experiments by Yashouv and Halevi (1967) failed to show growth heterosis, while other experiments on farms and elsewhere have shown heterosis to exist. Avault and Shell (1968) found that hybrid *S. niloticus (aureus?)* × *S. mossambicus* grew better than each of the parents, though allowance should be made for the fact that the hybrid population included a higher percentage of males (about 71%) which grow faster than females. Pruginin (FAO/UN, 1965) found that the cross between *S. niloticus* and *S. hornorum* grew somewhat better than either of the parent species. In a growth experiment with 865 fish/ha in ponds the following weights were reached after 200 days of rearing:

S. niloticus ♀	63 g
S. niloticus ♂	96 g
S. niloticus × *S. hornorum*	100 g

5.3.1 *Saratherodon mossambicus*

The shape of *S. mossambicus* fish is oblong, and the upper profile of the head is concave. The pectoral fins are as long or a little longer than the head, and the color of the body is olive gray, sometimes brownish or blackish, depending upon environmental conditions. During the breeding period females become grayish with black dots, while males become strikingly black, with the lower parts of their heads and mouths white and the edges of the caudal and dorsal fins distinctly red. *Sarotherodon mossambicus* originated on the east coast of Africa. In 1939 it was transferred to pond culture in Java, Indonesia and since then has been spread throughout Southeast Asia and South America.

Sarotherodon mossambicus is omnivorous. Fry feed on diatoms, planktonic algae, and small crustacea. Adults feed chiefly on algae but also on zooplankton, worms, insect larvae, insects, and detritus (Chimits, 1955).

The biggest drawback of this fish in ponds is its early reproduction. It attains sexual maturity at a length of 8–9 cm when about 2–3 months old (Chimits, 1955). The female produces 100–300 or more young fish per spawn and breeds throughout the year at intervals of about 30–40 days, except in subtropical climates where breeding is interrupted during the cold season. The lower limit of temperature is given as 9°C by Kelly (1957) and 8°C by Chimits (1957), though other sources give lethal temperatures up to 11–12°C (Allanson et al., 1971). This may be due to the length of time during which the fish is exposed to low temperatures. Chimits (1957) gives 14°C as the lethal low temperature by prolonged exposure of several days. Differences also can be due to geographical polymorphism and adaptation to different climates.

Sarotherodon mossambicus is very euryhaline and can tolerate high salinities. Many report on its ability to grow and even breed in sea water (Uchida and King, 1962; Popper and Lichatowich, 1975). Potts et al. (1967) have shown that young *S. mossambicus* can even live in 200% sea water.

Sarotherodon mossambicus is polymorphic with regard to growth rate. Fish under the same name and appearance may include different populations having different growth rate potentials. For instance, *S. mossambicus* from South Africa shows a much higher growth rate than populations from other areas in Africa and the Far East. While the South African fish can reach a weight of 5 kg in natural waters and 600 g in their first year in ponds (H. J. Schoonbee and T. Pike, personal communication), *S. mossambicus* from the Shire River basin in Malawi and Mozambique do not exceed 150 g either in natural water bodies or in ponds (Pruginin and Arad, 1977). Growth comparisons between *S. mossambicus* from South Africa and *S. aureus* from Israel have shown similar growth potentials. The weight attained by adult fish depends to a large extent on the population density in the pond.

The fish, if given a chance, can grow quite well and, according to Hora and Pillay (1962), can reach 850 g at the end of the first year. However, the fish is so prolific at such an early age and small size that the pond is soon filled with "wild" spawned fry, all competing for food in the pond; thus few reach a weight of 50–100 g. It is therefore strongly recommended that this species not be introduced into commercial fish farms.

5.3.2 Sarotherodon niloticus

Sarotherodon niloticus is a native African fish and has been reared in ponds since ancient Egyptian times. Murals and engravings dating from 1400 B.C. show the fish held in ornamental ponds.

Sarotherodon niloticus has a dark grayish coloration. The caudal fin has black narrow vertical stripes. The upper margin of the dorsal fin is black or gray. In breeding males, the ventral surface of the body and the anal, dorsal, and pelvic fins are black, and the head and body are flushed with red.

In spite of the identical appearance of fish within the species group, *S. niloticus* is genetically polymorphic. This can also be revealed by electrophoretic analyses (see Section 6.2.2).

In many African countries it has been found that *S. niloticus* seems to be growing at a faster rate than the other *Sarotherodon* species, both by itself or as a component in crosses with *S. hornorum* or *S. aureus*.

Sarotherodon niloticus feeds principally on phytoplankton (either in suspension or from the bottom), of which diatoms are an important item (Greenwood, 1958; Lowe, 1958). *S. niloticus* fry also feed on macrophytic detritus, rotifers and other zooplankton, insect larvae and water mites (Moriarty and Moriarty, 1973). It has been reported that in some African lakes this species feeds and digests blue green algae which is usually not digested by other fishes (Fish, 1955; Moriarty and Moriarty, 1973).

The lethal low temperature for *S. niloticus* is about 11–12°C (Chervinski and Lahav, 1976; Denzer, 1967; Bardach et al., 1972). Like all mouth brooders the number of eggs per spawn is small in comparison with most other pond fishes. Egg number can reach 1000–1500 in larger females (see Section 6.2.1), and it is nearly proportional to the body weight of the female. An important characteristic is the fact that sexual maturity in ponds is reached only at an age of 5–6 months. This gives the fish enough time to grow to market size before breeding begins (see Section 7.3.3).

5.3.3 *Sarotherodon aureus*

Sarotherodon aureus is a fish indigenous to Israel. Its natural habitat was the Hula Lake and the Hula Swamps which were drained about 20 years ago. It is similar in appearance to *S. niloticus* but can be distinguished from the latter by a number of morphological characteristics (Fishelson, 1962, 1966; Trewavas, 1965). The caudal fin of *S. aureus* is unmarked, has vague, irregular dark markings, or a more or less complete dark reticulum with white or colorless meshes (see Figure 5.4). The upper margin of the dorsal fin is pink. Differences between

FIGURE 5.4. *Sarotherodon aureus,* male. (Courtesy of S. Rothbard, Gan-Shmuel, Israel.)

the two species become clearer at the time of breeding, when the color of the body and head of *S. niloticus* becomes reddish with dark caudal and dorsal fin margins, while that of *S. aureus* is metallic blue with red caudal and dorsal fin margins. There are, of course, other meristic differences.

In nature, *S. aureus* feed on a wide array of organisms according to the availability of food in the specific water body. The list includes various species of phytoplankton, zooplankton, and zoobenthos (Spataru and Zorn, 1976). When introduced into fishponds, however, and under severe competition, such as that prevailing in polyculture ponds from other species of fish, *S. aureus* becomes mainly zoobenthophagous and detritophagous. Spataru (1976) examined the digestive tract of *S. aureus* from polyculture ponds. The most important items were oligochaetes (22–28%) and detritus (21–28%). It seems that in these conditions *S. aureus* feed on the ooze of the upper layer of the pond bottom. Spataru (1976) found that the stomach of *S. aureus* was not full, which can explain the positive response of *S. aureus* to supplementary feeding, provided an appropriate food is given.

Hybrids between *S. aureus* and *S. niloticus* seem to feed more on phytoplankton and zooplankton. An advantage of the hybrids is that they are caught much more easily by nets. *Sarotherodon aureus* avoids nets by burrowing into the mud underneath the lead line. It is therefore difficult to harvest and control its numbers by thinning.

The lowest lethal temperature for *S. aureus* is 8–9°C (Sarig, 1970; Chervinski and Lahav, 1976). This seems to be one of the lowest for the tilapia species and may be due to the adaptability of this species to the climate of its region, which is the most northern geographical distribution of the tilapia.

The breeding process and fecundity of *S. aureus* are described in Section 6.2.1. It should be noted here, however, that *S. aureus* has a relatively large number of eggs per spawn as compared to the other mouth breeding tilapias. This number naturally depends on the size of the female but can be up to about 2000 eggs. Fryer and Iles (1972) quote Liebman who counted 1300 embryros in the mouth of *S. aureus* from Lake Tiberias. Like *S. niloticus* it also reaches maturity only at an age of 5–6 months, which gives it an advantage in culture during the first part of the life cycle.

5.4 GRAY MULLET

Species of gray mullet are being cultured successfully in China, Hong Kong, the Philippines, Indonesia, India, Japan, Israel, and Italy (Lin, 1940; Hora and Pillay, 1962; Thomson, 1963; Huet, 1970). Mullet are

FIGURE 5.5. Gray mullet, *Mugil cephalus.* (Courtesy of Ray Coleman, Ma'agan Michael, Israel.)

catadromous fish which spawn in the sea but enter estuaries and the lower reaches of rivers or into lagoons where the water is brackish.

The fish have a cylindrical shape and have two dorsal fins. They have a terminal mouth with thin lips with no barbels or skin folds. The scales are cycloid and of medium size. They have prominent adipose eyelids which sometimes, as in *Mugil cephalus,* obscure the eye, leaving a narrow slot over the pupil (see Figure 5.5).

Though Thomson (1963) quotes Kesteven (1942) and others with respect to the occurrence of occasional hermaphroditic individuals, mullet are usually heterosexual. No external morphological characteristics are available to distinguish between males and females.

Mullet reach sexual maturity at an age of 2–5 years, depending upon temperature. The spawning season is from the end of summer until early winter. The fish then school and migrate to the deeper parts of the sea and spawn on the surface over a depth of about 150 m or more (Thomson, 1963). The number of eggs produced per female is 1–3 million. They are round, transparent, have a diameter of 0.93–0.95 mm, are straw colored, and have a large oil globule which provides buoyancy.

The eggs hatch after about 48 hr. The larval stage lasts for about 30 days, and the larvae are then about 20 mm in length. They school and are attracted to the estuaries by rheotaxis or chemotaxis. They enter the estuaries at a size of between 20 and 30 mm when they are about 2–3 months old.

The mullets have a very wide geographic distribution. The gray or striped mullet (*M. cephalus*) is found in the coastal waters, estuaries, and lagoons of tropical and subtropical zones, roughly between latitudes 42°N and 42°S where the average monthly water temperature usually does not drop below 16°C and summer temperatures are over 18°C.

Because of the euryhaline nature of some of the mullet species, especially *M. cephalus*, they can be cultured successfully in freshwater ponds either alone or, more often, in combination with other fish species in polyculture. *Mugil cephalus* is cultured in China, Hong Kong, India, Japan, and Israel. *M. tades* in India, Pakistan, and Indonesia; and *M. corsula* and *M. dussumieri* in India (Hora and Pillay, 1962). Six species of gray mullet are found on the Mediterranean coast of Israel. Only two of these are cultured in freshwater ponds: *Mugil cephalus* and *M. capito* (*Liza ramada*). The rate of growth and survival of the other four species (*M. auratus, M. saliens, M. chelo,* and *M. labeo*) are too low to stimulate commercial interest.

The growth and survival of *M. cephalus* in freshwater ponds are higher than those of *M. capito* (Perlmutter et al., 1957); therefore, the former is preferable as a pond fish. However, the number of fry of *M. cephalus* caught in the estuaries is not always sufficient. In Taiwan (Liao, 1974) and Hawaii (Kuo et al., 1974; Nash et al., 1974) spawning of *M. cephalus* has been successfully induced. Mullet fry production is not yet available on a commercial scale, however, and the culture of this species is still dependent on the collection of fry from nature.

Since insufficient *M. cephalus* fry are caught in nature, *M. capito* is also cultured in Israel. A degree of success has been achieved in the artificial reproduction of *M. capito* (Yashouv and Berner-Samsonov, 1970), but here again it has not reached the commercial stage. The fry of all species are collected in the estuaries. In Israel fry are collected during the autumn and winter, mainly following sea storms and rainy days (see Section 6.4.2). In India this is done during winter (December to March) (Hora and Pillay, 1962) and in the Philippines in the spring from April to July (Thomson, 1963).

Growth and survival of mullet in freshwater ponds depends, as with other fish, on the weight of the individual fish and their density. Pruginin et al. (1975) found the following growth rates for *M. cephalus* in ponds:

Ponds	Weight Range (g)	Density (fish/ha)	Growth Rate (g/day)
Nursing	68–200	6800	0.7
Rearing	50 to 415–530	75–135	2.9–3.7
Rearing	250 to 700–750	85–90	4.0–5.5

Young *M. cephalus*, up to about 35 mm total length, are carnivorous, feeding mainly on microcrustaceans (Odum, 1970; Zismann et al., 1975). Larger *M. cephalus* change their feeding habits and consume mainly microalgae and detritus (Thomson, 1963). Diatoms constitute a major part of the diet of young *M. cephalus*, while the amount of detritus in the digestive tract increases with the size of the fish (De Silva and Wijeyaratne, 1976). This seems to indicate that adult *M. cephalus* feed mainly on the ooze of the bottom of the pond and very little from the water column (Odum, 1970). *Mugil cephalus* seem to be able to strip off the microorganisms such as bacteria and protozoa that develop on detrital particles and utilize them as food while the detritus particle itself may be excreted. The large bacterial population in mullet guts also enables the fish to utilize nonprotein nitrogen (NPN) as a source of food, similar to what can be done by ruminants. Cowey and Sargent (1972) quote Vallet who compared the growth of gray mullets receiving a diet containing 4% nitrogen as protein (= 25% crude protein) with those receiving 2% protein nitrogen + 2% urea nitrogen. Vallet found that the urea could replace protein, at least in part. Leray (1971) reported on quantitative trials with mullet fingerlings which showed similar results.

Mugil capito feed mainly on plankton. Neither *M. capito* nor *M. cephalus* can feed on cereal grains, but they feed readily on protein-rich pellets or meals.

6

Breeding
and Nursing of Fry

6.1 CARP

6.1.1 Selection of Brood Stock

There is a natural tendency for fish farmers to select brood stock from among the largest fish and those which have grown fastest in the rearing ponds in the hope that these traits are additive and will be inherited. In Europe it was also customary to select for body size proportions (greater height/length ratio) and scale cover (mirror carp with a limited number of scales); both because of market demands and in the belief that these are correlated with a higher growth rate. Moav and Wohlfarth (1968) showed not only that there seems to be no correlation between body proportions and growth rate, but also that the selection of the fast growing fish in a population for brood stock, as suggested by Schäperclaus (1961) and Huet (1970), does not result in faster growing offspring (though selection for slower growing fish did lead to a noticeably reduced growth rate response in the offspring). These authors concluded that the carp population underwent a long continued natural and artificial selection for growth rate and, there-fore, exhausted their additive genetic variability and no longer re-spond to selection.

There is, moreover, a strong inbreeding depression in growth rate; that is, if the "selection" is done on the same population of brood stock and their offspring, the growth rate of the fish will tend to be lower and a large number of deformed fry will appear. It is clear that inbreeding should be avoided, and brood stock should be selected from sources as divergent as possible. Since no correlation was found between body proportions (height/length ratio) and growth rate, where there is no special demand in the market for a "high" shaped carp, there is no point in selection for that trait. Scale coverage, however, proved to be important. The carp cultured in most countries of the world are scaly, that is, fully covered with scales. However, in some countries, mainly in Europe and Israel, the mirror carp pattern of scales—a limited scale coverage under the dorsal fin and a few on the ventral side—predominates. Occasionally, however, other types of carp may be found in the ponds: leather carp with no scales at all and line scaled carp with a cover of scales mainly over the lateral lines. The mode of inheritance of these four scale patterns of carp is well known from Europe and was studied in Israel by Wohlfarth et al. (1962). The leather and line scaled carp carry a lethal gene that retards their growth. When this gene appears in homozygous individuals it will cause death of the offspring. Wohlfarth et al. (1962) confirmed that the leather and line scaled carp have inferior growth rates. Since breeding of these genotypes will result in lower survival and reduced growth rates, they should not be selected as brood stock.

Examination of the growth rate of different genetic lines of carp chosen at random from different farms in Israel and imported from abroad (Netherlands, Yugoslavia, and Taiwan) has shown considerable differences. Crossing these lines resulted in noticeable heterosis, with the crossbreds having higher growth rates than their parents (Wohlfarth and Moav, 1971). Similar results have been shown by Bakos (1979) in Hungary. The hybrids also contained less fat in their flesh. It is obvious that higher yields can be achieved by selecting brood stock in which the males and the females will be of different lines. This leads to the highest growth rate among progeny under the environmental conditions prevailing on the farm. This last point was found to be important since a considerable amount of interaction exists between the performance of crosses and environmental conditions (Moav et al., 1974). When the density of fish in a pond is low, and when supplementary feeds constitute an important part of the total food supply of the fish, the European lines, especially the crosses, perform best. On the other hand, when fish density is increased, and when cattle manure is used for increasing the production of natural food in the pond (see Chapter 9), the crosses between Chinese "big-belly" carp and the European or Israeli genotype prove to be hardier and can survive the harsh conditions which may develop in manured ponds. They also show better growth rates compared with the European genotypes (Moav et al., 1975; Wohlfarth et al., 1975). Similar results were also found by Suzuki (1979) on crosses between a race of Japanese common carp (Yamato carp) and two varieties of the European carp, the scaly and mirror carp. The Japanese common carp is probably related to the Chinese "big-belly" carp mentioned above. Suzuki found that in Japan (probably in conjunction with feeding protein-rich pellets), the European carp, especially the mirror carp, had a higher growth rate as well as better feed conversion efficiency than the Yamato carp. The survival rate of the mirror carp, however, was remarkably lower than that of the Yamato carp. The crossbreed of these two races showed a much higher growth rate than either of the two parents; the survival rate also was higher than that of both and the feed conversion rate was better.

Keeping "pure" genetic lines and testing their progenies for growth are difficult tasks that cannot be carried out on most commercial fish farms. They can be better handled by a research institute or specialized farm. In Israel this was done by the Fish & Aquaculture Research Station, Dor, together with five fish farms which became breeding farms supplying selected brood stocks and crossbred fry. Selected brood fish are branded with a hot wire (Moav et al., 1960) with a certain brand mark by which they are then also known. For instance, one of the best crosses developed at Dor is between the Yugoslav genotype (Π) and Dor 70 (W). Today, most of the farms in Israel, if not all, are using only selected lines according to anticipated environmental conditions during rearing.

Brood stock from selected lines should not be deformed or show body degeneration. They should be at least 1 year old. In practice, most of the brood stocks used are 2 years old and weigh about 2–3 kg. They can be used for a number of years until they reach a maximum size of about 8 kg. Larger fish are difficult to handle. It has been established that the size of the eggs and the resulting hatchlings are correlated with age and weight of the female (Alikunhi, 1966). The older and larger the female, the larger the eggs and hatchlings. Larger hatchlings have a better chance of survival. The number of eggs is also correlated with female size. The larger the female, the more eggs produced; 100,000 to 150,000 eggs are produced per kilogram of female's body weight. Due to losses during incubation and early nursing, however, only part of these survive. From a fish of 3–5 kg it is estimated that 50,000–200,000 fry of 0.5–1.0 g will be obtained, depending on environmental conditions.

6.1.2 Segregation of Brood Stock

The sex of carp can be distinguished externally, but only in the period before spawning. The gonads of the females are developing then, and this is recognized, externally, by swelling of the abdomen. In males the milt runs freely when the abdomen is stripped slightly towards the urogenital pore. At this stage it is advisable to segregate the sexes in order to prevent nonplanned spawning. Specimens not producing milt are assumed to be females. It is also advisable to tag or brand the fish with a permanent mark so as to distinguish among them, since this should not be done later during the spawning period. Tagging or branding enables fish to be held, if necessary, in rearing ponds together with market fish during most of the year.

The density of brood stock in the segregation pond depends upon their weight. Standing crop generally does not reach over 1.5 ton/ha. The segregation pond should be as rich as possible with natural food. In cases where the fish are fed supplementary feed, this feed should be rich in protein rather than in carbohydrates, so that the development of the gonads instead of the accumulation of fat will be stimulated.

The female segregation pond should be cleaned of grass and weeds, and the flow of fresh water into the pond should not be so large as to stimulate spawning, which may take place without the presence of males when environmental conditions are suitable. If brood stock introduced into the segregation ponds is composed of young fish which are being segregated for the first time, it is a good idea to recheck the sexing again in March. This is especially important in the female pond in order to verify that no males were introduced by mistake. This recheck is not necessary when the females are older and are already tagged or branded. In order to avoid unnecessary stress it is advisable not to disturb the brood stock (e.g. by netting) for at least 2 months before spawning.

After spawning the brood stock are returned to the segregation ponds. In order to save pond space and provide brood stock with better conditions, they can be stocked in a rearing pond together with fish reared for market. They must, however, be stocked into segregation ponds at least 3–4 months prior to the next spawning season.

6.1.3 Spawning Seasons

Spawning of carp in Israel takes place during two seasons:

1   The natural spawning season is in spring during April–May when the water temperature ranges from 18 to 24°C. As a rule spawning is carried out when the temperature becomes stable and there is no danger of a sudden drop which might affect hatching and survival of the fry. The eggs themselves are not harmed by the drop in temperature, but the hatching takes longer, leaving natural enemies more time to take a toll on the eggs and newly hatched fry. Survival rate is thus much lower.

2   In order to avoid "wild spawning" in the rearing ponds during the next spring, late spawning is carried out in August (see Section 7.4.1). This ensures that the fry will not reach sexual maturity before reaching market size. This is not the natural spawning season of carp in

Israel; therefore, special techniques should be employed. Spawners should be preselected, and those which are naturally inclined to be late spawners are chosen. If this process is repeated for a few years, it is possible to have natural spawning in August. Spawning must often be aided by a stimulating injection of carp pituitary for the females (see Section 6.1.5). The number of fry per female is much lower, only about 10,000 fry/kg, in the late spawning season compared to the spring season.

6.1.4   Preparation of the Spawning Pond

There is no restriction on the size of spawning ponds. Spawning can take place in a pond only several square meters in size or in a pond of 4–5 ha. When spawning is arranged in a small pond and the density of brood stock is higher than usual, fertilized eggs or young fry should be transferred to a larger pond for initial nursing to a size of about 1 g. When stocked at the regular density, or as is customary in some farms, at low density, the spawning pond itself serves as the initial fry nursing pond.

It is advisable to dry the spawning pond before spawning. Exposing the pond bottom to sunlight has an important sterilizing effect and prevents the rapid development of predators and parasites in the pond. When, after draining, pools remain in the pond, they should be sterilized with quicklime. If the pond has been dry for some time and weeds have been growing in it, these should be mostly removed, leaving some patches which may serve as spawning substrates.

In Israel the most common spawning substrate used for carp is made of needle tree branches (pine, casuarina, or cypress). The size of the spawning substrate for each female is about 10 m². The larger the spawning substrate the better are the chances for a larger number of eggs to be attached to it and not fall to the bottom of the pond (see Figure 6.1). Those which fall are usually not fertilized, and if they are, they become covered with silt and do not hatch. The spawning substrate is positioned so that it will be covered to a depth of about 10–20 cm of water when the pond is filled.

Instead of natural branches, artificial spawning mats can be used. These are made of synthetic material, similar to needle tree branches in shape. In Indonesia "kakabans" are used as spawning mats (Alikunhi, 1966). These are made of fibers of the indjuk plant (*Arenga pinnata*). The fibers are thoroughly cleaned, and a thin layer (1.2–1.5 m in length) is pressed longitudinally between two bamboo lathes 4.5 cm in width. The margins of the fibers are trimmed to leave a mat about 40–70 cm wide. A number of these are laid transversely on a long

FIGURE 6.1. A pond ready for carp spawning. Note the spawning substrate in the middle of the pond. Brood fish have been introduced in this case even before the pond was completely full.

bamboo pole set in the water with the fiber ends of adjoining kakabans touching each other and thus forming a large spawning mat. The area of kakabans per female is roughly the same as given above (about 10 m² per 2–3 kg female). The mats are advantageous from the point of view of convenience, especially when mats with the eggs are to be transferred to other ponds for hatching and nursing. It is easier to estimate the number of eggs spawned on mats than on natural substrates. Some culturists claim, however, that better spawning is obtained on natural substrates than on artificial mats.

Breeding ponds are filled with water just prior to the spawning. It is important that the water has not previously been used for holding fish (e.g., from another pond) since "fresh" water is the main trigger for the spawning of carp.

6.1.5 Spawning and Early Development of Fry

After the pond is filled with water, ripe spawners are chosen from selected brood stock and introduced. Ripeness is recognized in females by the swollen and soft abdomen and by the protuberance of the genital pore. In males it is recognized by the easy flow of milt after a gentle squeeze of the abdomen towards the genital pore. The weight of females is usually 2–5 kg each, with the males somewhat less. The numerical ratio between females and males should be 2:3, but since the males weigh less, the weight ratio is often 1:1. Spawners are introduced at densities of 10 females per hectare, or less. When more females are used per unit area, the fry have to be transferred to a nursing pond at a very small size or else, due to their high density, growth is retarded and the losses due to predation and parasitism are considerable. It is better, therefore, to introduce smaller numbers of spawners into the spawning pond in order to shorten the time during which the fry are vulnerable. Since the number of eggs per female in the late spawning season is much lower than during spring spawning, the number of spawners per unit area can be higher (up to 50 females and 70 males per hectare). In this late spawning season the spawners first get a stimulating pituitary injection (see Section 6.3.2). The females are injected with one pituitary per kilogram of body weight and the males with half of that dosage.

The fish usually spawn the morning after their introduction into the pond. The incubation period depends upon temperature; at 18°C it takes about 6 days for the eggs to hatch, while at 30°C only 2 days are required. With the pond temperatures prevailing in Israel during the spring (over 20°C), hatching takes place after about 48 hours. One day after spawning the extent of fertilization is estimated. The fertile eggs are transparent and the developing larvae can be seen, while infertile eggs become opaque and are later covered with fungus.

As mentioned above, if spawning is carried out in small ponds, the eggs can be transferred on the spawning substrate or mats to a separate, usually adjacent, hatching pond. The eggs would benefit from being covered with a wet cloth during transportation to protect them from wind and desiccation. The advantage of this method is in the separation between spawners and fry, as this decreases the chance of infestation of the fry by parasites which may be carried by the adults. The risk of fry being preyed upon by the spawners is also eliminated.

When spawning is carried out in a small pond and the eggs are not transferred to another pond for hatching, the spawning substrate is taken out 2–3 days after hatching. If possible, the spawners are also removed then by seining with a large mesh net. When spawning is carried out in a large pond, removal of the spawners is generally difficult and is done only when the pond is drained after nursing of the fry. Fry density plays an important role here. When the density is small, fry grow faster and pass the vulnerable time, up to 2–3 g, quickly and with relatively low mortality. At high fry density the losses are usually very high. There are various reasons for these losses, the most important of which are outlined below.

Parasites

1 *Dactylogyrus vastator* (and also the related monogenean trematodes *D. anchoratus* and *D. extensus*) attack crowded fry a few days after hatching up to a size of 2–3 g in spring (April–July). They can cause total mortality within 8–10 days after hatching (Paperna, 1963; Sarig, 1971). Fry over 20–35 mm (2–3 g), even when attacked, do not suffer much due to a rapid regeneration of gill tissues. This is also true for fast growing fry weighing less than 2–3 g in populations which are not too crowded.

2 *Gyrodactylus* spp. is another monogenean tremadode which may appear in masses in spawning and nursing ponds, though less frequently than the previous one and the damage is less extensive.

Prophylactic treatment is used on these parasites. Three days after fry hatching, Bromex (dimethyl 1, 2-dichlorethyl phosphate) is added to the pond at a rate of 0.2 ppm. If the pond is not treated and parasites subsequently appear, Bromex should be immediately added.

Aquatic Insects

Among the predaceous aquatic insects, the main pests are species of *Notonecta* which prey on fish eggs and young fry. In Israel they are especially dangerous during late season carp spawning (in August) and in the nursing of the silver carp and grass carp in June to August, since in summer the insect populations reach their peak. In warmer climates, however, the insects constitute a hazard all year round. According to FAO/UN (1965), water boatmen (Corixidae) and backswimmers (Notonectidae) were major pests in fry ponds in Uganda and preyed on fish larvae from hatching until the fry attained a length of 15 mm.

Control of insects during late season spawning as well as for nursing of silver carp is essential for success. This was previously done by applying 0.1–0.2 ppm of BHC (Lindane) to the pond water just before fish hatching (Lahav, 1971). This treatment was effective against the bugs for 7–10 days. The insects developed an immunity to this chemical, however, and treatment was changed to an application of 0.5 ppm (500 ml to 1000 m^3 of pond water) of Malathion (M-50), repeated after 24 hours. In late season spawning of carp, an additional treatment is often necessary. The main problem with these treatments is that the safety range between the dose affecting the bugs and that affecting the fish is small, and a minor error may result in fry mortality.

Another treatment used by some fish farmers is the application of either edible oil or mineral oil, at a dose of 40–50 l/ha, when there is no wind. The oil cover, even if temporary (20–30 minutes), causes death of the bugs which cannot surface for breathing. This treatment should be repeated every 2–3 days until the fry reach a size where they are safe from predation (about 15 mm in length). This takes about 10 days from the day of hatching.

Three days after hatching, when the yolk sack has been absorbed, feeding is started with a very fine feed, such as soft white cheese and/or sifted fish meal and oil meals. A few handfuls of feed are distributed along the banks of the pond.

6.1.6 Induced Spawning

Induced spawning of carp in the hatchery is not a common practice since it requires professional skill and appropriate hatchery installations. It has, however, a number of advantages over natural spawning, especially in the late spawning season. These include:

1 When spawning is carried out indoors the newly hatched fry are held in controlled conditions for about 10 days until they reach a length of 11–12 mm. Stocking them at this size into nursing ponds reduces the losses due to predation by insects during the nursing of the fry.

2 Shortening the outdoor nursing period lets the pond be used for increasing production.

3 Fry obtained by induced spawning are much more uniform in size compared to those produced by late season outdoor spawning. In tropical regions insect populations are more or less prevalent throughout the year, and the loss of carp hatchlings is very high. Under these conditions it may prove advantageous to produce fry under controlled conditions in an indoor hatchery and nurse them to a size which is less susceptible to predation.

The number of spawners required to produce a given number of fry will differ from spring season to late season spawning. The variation is related to differences in the number of eggs spawned per female and the percentage hatched. In spring, the spawn constitutes about 20% of the ripe female's weight, while in August it is only 10–12%. The fertilization and hatching rates are also higher in spring, reaching about 50%, as compared with only about 30% in the late season spawning. This means that in spring a female of 5 kg produces about 1 kg or 700,000 eggs, of which about 350,000 will hatch. Taking a mortality rate of fry during the early nursing period indoors and a secondary nursing period in ponds at about 50%, then 120,000 fry weighing 2 g each can be produced by one female. In August a female of 5 kg produces about 0.5 kg or about 350,000 eggs, of which only 120,000 will hatch. Assuming the same losses during the nursing period, about

60,000 fry of 2 g each can be produced from one female. Since not all the females will respond to injections the number of females which should be brought into the hatchery is doubled. Thus, in order to produce 1 million fry of 2 g each, about 16 females are required in spring and about 33 females in late season.

The number of males required is about two-thirds the number of females. Note that in natural spawning, the ratio between males and females is reversed.

Each female requires a separate holding tank of about 0.7 m³. The males are held in a common tank, five males to 1 m³. Since both tank space and the number of incubating funnels (see below) may be limited, spawning can be done successively every 4–5 days, according to available facilities. Remember, however, that the time period for spawning is sometimes limited, as is the case for late season spawning in Israel. Spawning before the end of July will not prevent "wild" spawning in the rearing ponds the next spring, while spawning later than about the middle of August will not leave enough time for nursing the fry to a size large enough to pass the winter safely. This leaves a limited period during which indoor facilities can be used, and their size should be planned accordingly.

Immediately after the transfer of the brood stock from the segregation pond to the hatchery, the fish are disinfected in a bath of 40 ppm Formalin for 2 hours. The dosage and sequence of pituitary injections may vary slightly in various hatcheries and in accordance with the ripeness of the spawners. The technique of extracting the pituitary glands is outlined in Section 6.3.2. A tested procedure is the injection of 0.5 pituitary per kilogram of female spawner 1 day after transferring the spawners into the indoor hatchery tanks. This 24 hour period is required for acclimatization of the spawners to hatchery conditions. Eight hours later, a second injection of 0.8 pituitary per kilogram of female is given. Spawning is carried out by means of stripping about 8–10 hours after the second injection. The males are injected only once, 24 hours after introduction into the hatchery tanks, with 0.5–0.6 pituitary per kilogram. The total number of pituitaries required to produce 1 million fry will be about 130 in the spring and 270 in August.

The female which is ready to spawn about 8 hours after the last injection shows clear signs of restlessness and swims in circles in the tank. When the time of spawning approaches, the ovipore of the female becomes swollen. The female is then gently transferred into a container of about 30 l containing an anesthetic solution of ethyl-M-aminobenzoate at a concentration of 100 ppm. After 3–5 minutes the female is completely anesthetized. A male is treated simultaneously. The anesthetic solution can be used repeatedly. After anesthetizing both the female and the male, they are dried thoroughly with a dry towel. This is important so as not to cause clumping of the eggs. The female is then gently stripped from the head toward the tail. The eggs should flow easily into a previously prepared bowl of about 5 l. Milt is similarly stripped into the same bowl, and the contents are immediately mixed. For fertilizing 1 l of eggs, 2–3 ml of milt are required.

The stickiness of carp eggs may impair proper fertilization and development as the eggs tend to clump together. Woynarovich (1962) has shown that this stickiness can be eliminated by treating the fertilized eggs with a solution of 0.4% iodine-free salt (NaCl) and 0.3% urea. The eggs and milt are first mixed gently for about 3 minutes. Then the solution is added, and the eggs are mixed continuously with the solution for about 20 minutes. The eggs are then allowed to settle, and the solution is decanted without exposing the eggs to air. These washings are continued, with decanting at 5 minute intervals, for about 10 times. In all, this takes about an hour. By this time the eggs are swollen to a diameter of about 2 mm, and the first cleavage occurs. The eggs are then washed with a 0.05% tannic acid solution for 20 seconds. A total volume of 3 l/kg of spawn is required for this purpose. These washings are repeated about five times, each successive solution being weaker by 0.01%. The concentration required for each successive washing is obtained by adding water to the stock solution. The eggs are then washed with fresh water for 5 minutes and transferred to the incubation funnel.

The incubation funnel is usually made of plexiglas (Figures 6.2 and 6.3). It is 60 cm in diameter at the upper rim and 80 cm deep. The

FIGURE 6.2. Large "homemade" incubation jars, made of transparent plastic sheets and a conical bottom. Each jar contains a batch of fish eggs. (Courtesy of S. Rothbard, Gan-Shmuel, Israel.)

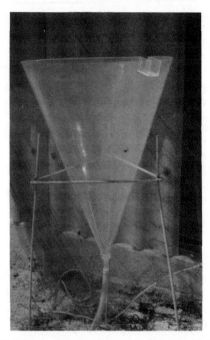

FIGURE 6.3.   Incubation funnel made of plexiglas.

bottom inlet has a diameter of 1.25 cm (0.5 in.). Water is introduced from the bottom and overflows through an outlet at the top into a tank also used to receive hatched larvae. Each incubation funnel can hold 250 g of spawn (about 175,000 eggs). At a temperature of 24–26°C hatching takes place after about 48 hours. The shorter the incubation time the better since losses are reduced. The larvae then pass by swimming through the outlet into the receiving tank.

The nursing tanks are about 0.5 m deep. The area required for the indoor nursing is about 1 m² per 100,000 fry. For production of 1 million fry (of 2 g each) about 20 m² are required. Water is exchanged at a rate of about 1 l/min per 1 m² of tanks. The tanks are aerated (one aeration stone for each 2.5 m²) and well illuminated with a lamp of 500 W, 1 m over the water surface, for 16 hr/day.

Feeding is done by automatic feeders operated by alarm clock mechanisms (see Figure 6.4). There is one feeder for each 1.25 m² of tanks. Commercial feed such as trout starter is not sufficient to maintain the fry for more than 1 week. Beyond this time the fry die rapidly. To extend this period to the 10–11 days required to bring the fry to a size of 10–11 mm, the food should either be organisms collected in the ponds or brine shrimp (*Artemia salina*) produced in the hatchery.

FIGURE 6.4. A "homemade" automatic feeder for fry, operated by an alarm clock. (Courtesy of S. Rothbard, Gan-Shmuel, Israel.)

Recently a yeast, *Candida lipolytica,* has been used successfully as fry feed. This feed gave better results than *Artemia* (Appelbaum, 1977; Appelbaum and Dor, 1978). Feeding starts about 48 hours after hatching and when the yolk sac is completely absorbed. Feed is administered continuously 16 hr/day. The total amount of feed required is 500 g per 100,000 larvae every 8 days.

6.1.7 Nursing of Carp Fry

The nursing of carp fry has several purposes, the most important of which are the following:

1 To rear the fry as fast as possible to a size where they are relatively safe from predation and parasites.

2 To bring the fry to a size which, when stocked in a rearing pond, will better use the productivity of the pond. This size is considered to be about 100 g (see Section 7.1).

3 To better use the pond area by stocking fry at higher densities when they are small and require less natural food and space.

These objectives are achieved by nursing the fry in a number of stages and sometimes by different methods. Initial nursing is done in spawning ponds where the fry reach a size of 0.2–5 g, according to density: the higher the density, the smaller their final weight. In late season spawning, the spawning substrate and spawners are removed about a week after fry hatching, and the initial nursing is continued for about an additional 14 days. At a density of 500,000 fry/ha, their weight at the end of this period may average about 0.5 g. When induced spawning is practiced, fry are held in the hatchery for about 10–11 days and only then are they transferred to an initial nursing pond. Density in this case is higher, 1.0–1.5 million/ha, and after 25 days the fry can reach a weight of about 2 g. This is due to the additional nursing and development in the hatchery as a result of which the fry start the nursing stage at a larger, more active stage. Anticipated losses may run 50%. If fry are stocked smaller than 10–11 mm in length, the pond should be treated against bugs with 40–50 l of mineral oil per hectare applied every 2 days until the fry reach the safe size. In order to enhance fry growth, the pond should be fertilized with chemical fertilizers (60 kg of superphosphate and 60 kg of ammonium sulfate per hectare) and manured with chicken manure (about 100 kg/ha) every 2 weeks.

From the spawning pond the fry are transferred to other ponds for the second nursing stage. The stocking densities in these ponds will depend upon the planned final weight of the fish. If the fish are to reach a weight of 10–15 g, the density should be about 100,000/ha, and if the aim is 50 g fish, about 20,000/ha should be stocked. For 100 g fish, stocking would be about 10,000/ha. In all cases the final standing crop is about 1 ton/ha. In the late season spawning technique, fry should reach as large a weight as possible before winter to reduce weather losses. During the period remaining before winter, fry can reach a weight of about 20–50 g. For this to be realized they should be stocked into the second stage nursing pond at a density not greater than 20,000 fry/ha (Peleg, 1971).

In the second stage nursing pond, the fry are first fed with ground cereal grains (sorghum, wheat, etc.) and at a later stage (after they reach 10 g) with whole cereal grains. Feeding with protein-rich pellets, used for carp, does not produce better results. Density is the main factor affecting growth rate (Hepher, 1978).

After nursing, the fingerlings either are transferred to rearing ponds until they reach market size, or they are concentrated in fingerling holding ponds at a high density (standing crop of 10 ton/ha and more) in order to be held for stocking at a later time. Due to high densities in the holding ponds, the fish may become stunted and not grow further, but this does not seem to affect their later growth in rearing ponds.

On some fish farms fry are nursed to a weight of only 10–15 g. They are then introduced into rearing ponds together with another carp population of larger fish that weigh over 250 g each. This is called *mixed nursing*. The density of fry in this method is that customary for the larger fish plus 20%—the expected loss. When the pond is drained for harvesting the larger fish are removed for market, leaving the smaller fish that have now attained a weight of about 200–250 g for restocking. When the latter are restocked in rearing ponds, another batch of 10–15 fingerlings is added together with them. In this way nursing pond area is reduced. The disadvantage is the need to sort fish during harvesting.

During the second stage of nursing, the major cause of loss is predation. Main predators at this stage are:

Water Snakes.   Water snakes (*Natrix* sp.) prey on considerable numbers of fry of a size smaller than 20 g. One snake's stomach may contain tens of fry. Snakes from all around the farm concentrate at the nursing pond and may cause considerable damage. They can be controlled with simple cylindrical traps (Figure 6.5), a number of which are set in the ponds near the feeding place and near the shores. The trap is made of 0.6 cm mesh wire screen formed into a cylinder. The diameter of the cylinder is 25 cm and its length is about 70 cm. It is fitted with a conical entrance at each end, one of which can be removed. The traps should be checked daily and the trapped snakes removed.

Water Fowl.   Most water fowl are not considered voracious fish predators, with the exception of three: pelicans, cormorants, and herons. The damage caused by the first two birds is not limited to nursing ponds since pelicans and cormorants can swallow large fish. Herons, on the other hand, cannot swallow large fish so most of their damage is caused in the nursing ponds when water levels are too low. According to Sarig (1971), more than 80 fingerlings each weighing 10 g were found in the gullet of a single heron. In order to decrease this damage,

FIGURE 6.5.   Snake trap for fishponds.

water depth in nursing ponds must be at least 70–80 cm at the shallow end.

Many methods of driving birds away, such as making explosive noises, are not always effective because the birds quickly learn and get used to these noises. Greater success has been achieved by extending wires across the length and width of the ponds. The best way is to drive the birds away with guns. The majority of seagull species prey on fish only when the latter are weak, sick, or come to the surface during anoxia. The appearance of seagulls sometimes serves as an indication of an unhealthy state of fish in the pond.

Frogs. Frogs may be serious pests for fry. Pruginin, working in Uganda (FAO/UN, 1965; FAO, 1967), has found that the frog *Xenopus* was exceedingly abundant in carp hatching ponds and fed on fry ranging from 20 mm (0.25 g) to 35–40 mm (2 g) in length. Another unidentified species of frog, present in lesser numbers, fed on fry up to the length of 70–80 mm (10–12 g). Experiments showed that these frogs were responsible for 86% of fry mortality.

In Israel *Rana* sp. prey on fry up to about 10 g, especially when the fish are concentrated near the banks. The damage caused by this frog has not yet been fully ascertained, but it may not be heavy since the number of frogs is limited.

Where frogs are abundant it may be advisable to construct a barrier 45–60 cm high around the nursing ponds to prevent intrusion. Pruginin and Arad (1977) report that such a barrier, made of a low brick wall topped by an angled sheet of corrugated galvanized steel, reduced the frog count from 300 per 0.1 ha pond to only 4. The survival of tilapia fry (*Sarotherodon mossambicus*) increased from 10,000 in unfenced ponds to 19,300–67,400 in the fenced ponds.

6.2 TILAPIA

The tilapia species which perform best in fish culture (*S. niloticus* and *S. aureus*) reach sexual maturity at an age of 5–6 months irrespective of size. Optimal temperature for spawning is 26–30°C. *Sarotherodon aureus* begin spawning at 20–22°C, while *S. niloticus* start spawning when the temperature reaches 24°C.

Most of the tilapias cultured in ponds are mouth breeders (*T. zillii*, which is a "guarder," is not considered to be a good fish for culture in ponds and, in some cases, is considered to be a pest). The breeding process starts when the male occupies a particular territory, digs a craterlike spawning nest, and guards his territory. The ripe female spawns in the nest, and immediately after fertilization by the male she collects the eggs into her mouth and moves off. The female incubates the eggs in her mouth for about 2 weeks and broods the fry after hatching. The young fry, after leaving the mouth, may find refuge in it again should danger threaten. As the young continue to grow, however, fewer and fewer are recovered by the female, until brooding finally ceases. The male remains in his territory, guarding the nest. He is able to fertilize eggs from a succession of females. It is obvious, therefore, that the number of females in the spawning pond should exceed that of the males, and that the number of fry produced will depend on the number of females stocked.

The total number of fry per female during a season will depend on the number of eggs she has in each spawn and on the frequency of spawning. According to Lowe McConnel (1955), the number of eggs per spawn may differ among species, but in any one species the larger the female the greater the number of eggs she can spawn. A female *S. niloticus* weighing 100 g produces about 100 eggs per spawn, while a female of the same species weighing 600–1000 g can spawn 1000–1500 eggs. A female *S. aureus* weighing 1000 g can produce about 2000 eggs in one spawning. It is clear that in order to obtain larger numbers of fry, it is better to select large spawners.

In regions with distinct thermal seasons (winter and summer), spawning will be limited to the summer. Since *S. aureus* start spawning at a lower temperature than *S. niloticus,* the spawning season is longer for the former species which can spawn three to four times during the summer, so that the total number of eggs produced per female can reach about 8000. The number of fry which can, in fact, be cultured is much smaller. Only the fry of the early spawnings have a chance to grow to market size (2–300 g) within the same year. Even if fry are kept over winter, for rearing the next year, there are some size limitations. The usual practice of second year tilapia culture is to use a monosex male population while the slower growing females are discarded after sexing (see Section 7.3.3). Since in most cases the wintering of tilapia is difficult and demands effort and expenditure, there is no point in holding females over winter. It is more economical to sex tilapia before winter and keep only the males. Sexing can be done on fish larger than 40–50 g with a reasonable degree of confidence. This means that fry should be nursed to this size. Since about 2 months should be allowed for nursing, only fry of the first two or three spawnings can be used. These generally yield about 3000 fry per female, which determines, to a large extent, the size of the brood population. These considerations do not apply in the tropics, where spawning is continuous.

6.2.1 Spawning in Ponds

Since the desirable tilapias are mouth brooders, there is no need for special arrangements in the spawning pond except for careful drying of the pond prior to spawning in order to eradicate unwanted fish. Care should be taken to prevent wild fish from entering the spawning pond during refilling. The major tilapia species cultured in ponds spawn at a depth of not over 50–60 cm. It is therefore desirable to use, for tilapia spawning, a pond with only a slight bottom slope and to fill it only to the depth indicated above. In a pond having greater depth, all the spawning nests are concentrated in a narrow strip near the banks of the pond at depths of less than about 60 cm; thus the total area for breeding becomes limited.

Optimum density of spawners is related to the fact that during the breeding season male tilapia are territorial and each requires a certain amount of space. The density of males is usually 100–250/ha, depending on female size. When the females are large fewer are stocked per hectare and fewer males are required. It seems that a greater number of males does not necessarily result in a larger number of fry per unit area. Because, as explained above, the mouth brooding tilapias are not monogamic, the number of females should be greater than that of males by a ratio of 3:1 or 4:1. If, however, the males and females are of different species, more males are needed for hybridization, and the ratio should be 1:1.

Tilapia do not stir up the water and make it turbid as do the common carp, which are bottom burrowers. Therefore, unless naturally turbid, tilapia pond water remains clear, and light penetrates to the bottom of the pond. As a result, filamentous algae often develop on the bottom of the pond. This creates difficulties in harvesting small fry from the pond since the algae and fry become entangled in the net and are difficult to separate (for other hazards of filamentous algae see Section 8.5). In order to avoid this problem, male carp averaging about 500 g (250–300/ha) can be introduced into the tilapia spawning pond. They stir up the water and reduce light penetration, thus preventing the development of filamentous algae. The carp should be of one sex only so that they will not spawn and cause the additional work of sorting out the two species.

Tilapia females do not spawn simultaneously, and so the fry are not all of the same size. The harvesting of fry is therefore a continuous process. They should be removed from the spawning pond when they reach a weight of 0.5 g. This is done by seining with a drag net of 4–5 mm mesh. This is repeated every 10–14 days. Fry in the spawning pond are fed a small amount of pelleted feed. Unlike carp, tilapia do not respond to feeding with cereal grains.

6.2.2 Hybridization

Hybridization of two species of tilapia in order to obtain an all-male hybrid population was started by Hickling (1960, 1968). He crossed what looked to him like two subspecies of *S. mossambicus,* a female from Malaysia (Malacca) and a male from Zanzibar, and got a 100% male offspring population. Later, Trewavas (1968) found that the Zanzibar "subspecies" was really a distinct species—*S. hornorum.* Pruginin (FAO/UN, 1965; FAO, 1967) crossed a number of different tilapia species from various parts of Africa and Israel for the same purpose. Almost all his crosses produced a high percentage of males and several—*Sarotherodon niloticus* or *S. nigrus* × *S. hornorum, S. niloticus* × *S. variabilis,* and *S. niloticus* × *S. aureus*—resulted in all-male progeny.

The phenomenon of abnormal proportions between the sexes of hybrids of different tilapia species has been studied by Chen (1969) and Jalabert et al. (1971). The reader is referred to those papers as well as to that of Avtalion and Hammerman (1978). They present a number of theories to explain the mechanism, though none of these has yet succeeded in explaining it fully. Many of the crosses which give all male hybrids involve *S. hornorum* as the male parent. The cross between *S. hornorum* and *S. niloticus* has an additional advantage in that the offspring are different in appearance from either parent. This can help prevent errors in identification and the use of the crossbreed for reproduction.

Two hybrid combinations have been tested on a large commercial scale in Israel. One is between female *S. vulcani* (native to Lake Rudolph) and male *S. aureus,* and the second is between female *S. niloticus* and male *S. aureus.* The first combination usually results in 70–85% male offspring. This does not eliminate the need for manual sexing, but it does reduce the number of female fry which are destroyed, thus lowering the cost of sexing per each male fry.

The second combination (*S. niloticus* × *S. aureus*) was made by Pruginin in Uganda and resulted in 100% males, but when it was repeated in Israel with *S. niloticus* introduced from Uganda, (Kajansi Experimental Station), it did not produce the same results. Crossbreeding of some individual fish gave batches of 100% male fry, but this trait was not stable with the same fish nor was it inherited by the offspring of the parental lines. It was clear that in spite of external appearance and other characteristics used in the identification of these tilapias (according to which they were all *S. niloticus*), the fish were not genetically homogeneous. As an additional means of determining the genetic homogeneity of the species, electrophoresis of the serum was applied. The analyses showed that the *S. niloticus* brought from Uganda were not homogeneous but rather were biochemically polymorphic and lacked a definite electropherogram. For comparison, another group of *S. niloticus* was imported, this time from a natural habitat in Ghana. The electropherograms of this group were much more homogeneous, and crossbreeding in commercial size ponds resulted in 90–100% male fry. The electropherogram pattern of this group, therefore, served as a genetic marker for selecting spawners. The serum of all fish used for crossbreeding is first analyzed electrophoretically, and only those showing the same pattern are selected. A hybrid combination that produces 100% males is not available on a commercial scale as yet, but even a population of 95% males eliminates the need for manual sexing, since 5% females in the population will not produce so many fry that stunting will occur.

Producing a monosex male population is not the only advantage crossbreeds have over pure species. The following are some other advantages:

Better Catchability. *Sarotherodon aureus* is difficult to catch by seining. It burrows into the mud, slips below the lead line, and evades the net. This makes it hard to harvest without draining the pond and almost entirely prevents thinning of the population during the growing season. Crossbreeds of *S. aureus* and *S. niloticus* behave like the latter—they do not evade the net and are easily caught.

Efficient Use of Trophic Niches. The feeding behavior of the crossbreeds is similar to that of *S. niloticus*. While *S. aureus* feeds to a large extent on the detritus and the ooze at the upper layer of the bottom, *S. niloticus* and the hybrid feed mainly on phytoplankton from the water column. This can be of importance in polyculture where an effort is made to utilize all of the trophic niches available in the pond.

Tolerance of Low Temperatures. *Sarotherodon aureus* is more tolerant of low temperatures than *S. niloticus*. While the former can withstand temperatures as low as 9°C, the latter's limit is 11–12°C. The crossbreed will tolerate about 10°C. These differences, though seemingly small, may be critical in some areas.

Growth Rate. As discussed above, some claim that the crossbreeds do not show heterosis, while others claim that heterosis does exist and that the crossbreeds grow about 25–30% better than either parent species (see Section 5.3).

Marketable Color. The color of the hybrid *S. aureus* × *S. niloticus* is much lighter than crossbreeds *S. aureus* × *S. vulcani*. The latter are almost black. The lighter color is preferred in the market.

Commercial crossbreeding is carried out in ponds, as described above. Drying of spawning ponds prior to the spawning is very important to eliminate the presence of other species of tilapia. Strict measures should also be taken to avoid fish intrusion from outside. The fish should be reexamined with regard to their sex to make sure that no errors have been made. The ratio of males to females in hybrid breeding is 1:1. In order to be absolutely sure of species identification, the fish should be hatchery bred and marked according to species (see Section 6.2.3). The fish should not be kept in the spawning pond for more than 3 months. A longer period may result in backcrosses between the offspring and their parents.

6.2.3 Hatchery Breeding

Monospecific breeding of tilapia in the hatchery is done to obtain brood stock of known species and known source for use in hybridization. Advantages of breeding in a hatchery are that controlled environmental conditions can easily be maintained and spawning will continue year round rather than only in the summer. An optimum temperature range of 25–29°C is maintained, breeding aquaria are lit 12–14 hr/day (by means of fluorescent bulbs suspended about 50 cm above the surface of the water), the water is kept clear and is replaced every 2–3 weeks at which time the aquaria are thoroughly cleaned, the water is aerated constantly, and spawners are fed a diet containing 38–40% protein and vitamins. In cases where fish experience severe injury or excess scale loss due to handling, the aquarium water is treated with malachite green as a prophylactic against *Saprolegnia* infection. Under the specified conditions females spawn readily and incubate eggs in the normal fashion. The light regime is important in determining the hour of spawning. The females are ready to spawn between the tenth and twelfth hour of light or, sometimes, spawning is delayed until the early hours of the following morning. If, for reasons of convenience, it is desired that spawning take place at a certain hour of the day, the light can be manipulated accordingly.

Two methods of breeding have been used: (1) natural fertilization in aquaria and (2) stripping and artificial fertilization. Both methods are described in detail by Rothbard and Pruginin (1975).

The first technique is relatively simple. Spawners are selected according to external characteristics and their electrophoretic patterns (see Section 6.2.2). Selected fish are then introduced into a long aquarium (200 × 50 × 40 cm). The aquarium is long enough to enable the male to establish a territory for the breeding nest. A "family" is established in the aquarium. This consists of one male and seven to ten females. The "family" is created at an age of 4–5 months before the fish reach sexual maturity and at a weight not exceeding 100 g. When larger fish, or more than one male, are introduced into the aquarium, they may become very aggressive—the male toward the females and the females amongst themselves. Mutual attacks frequently result in mortality.

When the fish reach sexual maturity their colors become very distinctive, especially in the males. *Sarotherodon aureus* becomes bright orange or red on the fringes of the dorsal fin; *S. niloticus* becomes black on the tips and fringes of the dorsal and caudal fins. The male digs a nest in the fine gravel on the bottom, at one corner of the aquarium. If the bottom is bare and made of glass, however, or any other hard material, the male simulates the nest digging, with the same movements as if the bottom were soft. The male drives off all females except the ripest one. Courtship may last for several days, but spawning and fertilization require less than 2 hours. Spawning and fertilization proceed even on bare glass or any other hard bottom. Immediately after spawning and the collection of the eggs into her mouth, the female leaves the nest area. The male is then ready for courtship and mating with the next ripe female. Daily inspection should be made to see which females have spawned. The eggs are left in the female's mouth for 3–5 days and then taken out and incubation continued artificially, in containers placed on a shaking platform (Figure 6.6) or a "Zuger" incubator. The shaking platform keeps the eggs

FIGURE 6.6. Shaking platform for the incubation of tilapia eggs. (Courtesy of S. Rothbard, Gan-Shmuel, Israel.)

moving continuously, preventing sticking and fungus infestation. At temperatures of 25–27°C the eggs hatch about 50 hours after transfer to the containers, but the larvae are kept on the platform for another 8–10 days until the yolk sac is completely absorbed. They are then transferred to larger tanks or to cages placed in ponds for nursing. During incubation, dead eggs are removed or else they can spoil the whole batch. If turbidity occurs because of egg decomposition all the water should be changed.

Removal of the eggs from the female's mouth is necessary since overcrowding may lead to cannibalism of the fry. Females from which eggs are taken can be ready for spawning within 1–2 weeks, much sooner than those which incubate and brood their fry. Thus it is possible to induce a female to breed 10 to 12 times a year, as compared with 3 to 4 times a year in ponds, and about 5 times a year in the hatchery if females brood their young. The number of fry obtained from each female annually is, therefore, considerably increased.

Induced breeding, described in detail by Pruginin and Cirlin (1973), and by Rotbard and Pruginin (1975), is important when crossbreeding is desired. In the hatchery the fish do not readily hybridize. Stripping techniques must be employed, and the entire incubation is carried out on the shaking table. There is no need for pituitary injections to stimulate egg development. Ripe spawners are chosen from their own, monospecies families when they are ready for natural spawning according to the following external characteristics:

1 Intense display of pigmentation—this is particularly conspicuous and easy to identify in males;

2 Swollen genital papilla; and

3 Erected scales caused by the stretching and swelling of the body during courting.

Eggs of the ripe females are stripped into a container, and then milt from the male of the second species is stripped into the same container. Eggs and sperm are gently mixed with a feather for 2 minutes, after which 10 ml of saline solution are added. Eggs and sperm are mixed for 2 more minutes. They are then rinsed with a small flow of tap water to remove remaining sperm and other tissues and transferred to containers on the shaking platform.

6.2.4 Sex Reversal

Another recently developed method of obtaining a monosex male population is sex reversal (Shelton et al., 1978). This technique has not yet been introduced on a commercial scale. Two traits related to sex determination in tilapia enable the technique of sex reversal:

1 Sex is determined at a relatively late stage in fry development—during the 3–4 weeks after hatching—when the fry are below 18–20 mm in length.

2 Sex is quite labile shortly after hatching and can be affected by internal and external factors.

It was found that administering androgenic hormones during this critical period can reverse the entire or at least the majority of the fry population into effective males. This has been tried in different ways, such as submerging the fry in an aqueous hormone solution (Eckstein and Spira, 1965) or injecting hormones, but the most convenient and effective method is oral administration. Hormones are incorporated in the fry feed (Guerrero, 1975). There are many androgens that can be used, but the most active when given orally are ethynyltestosterone (ET) and methyltestosterone (MT).

Since it is important that no natural food that might be preferentially eaten be available to the fry, the treatment can be carried out only in tanks, preferably indoors. The area of required tanks may impose some constraints when commercial scale sex reversal is attempted. According to Shelton et al. (1978) the growth of fry in tanks, and thus the period of hormone treatment, depends on density and temperature. At the highest density tested—2600 fry/m²—it took 4 weeks to grow 62% of the fry to a size of 18 mm, which is considered to be the target size at which sex reversal is complete. At a density of 160 fry/m², 100% of the fry passed this target size in 4 weeks. Thus, an additional 1 or 2 weeks are required for the entire population to reach 18 mm at the higher density. If we consider that two fry must be produced for every tilapia fingerling stocked in the rearing pond (see Section 7.4.1) and assume a stocking density of 3000 fingerlings/ha, then 2.3 m² of indoor tank area will be required for a duration of 5–6 weeks for each hectare of rearing ponds, assuming a tank density of 2600 fry/m² during sex reversal. This is quite a large area and will be reflected in the price of the fingerlings.

In order to treat fry for sex reversal, large numbers of newly hatched fry of more or less uniform age and size must be ensured. This can be done by stocking a small pond with tilapia brood stock at relatively high density. Every 14 days the water level in the pond is lowered and the spawners seined. Due to the stress associated with netting, the females eject the fry from their mouths. These fry school together and swim in the upper layer of the water. There they can be scooped up with a dip net of a "mosquito" mesh size. Fry are then transferred to the treatment tanks.

Hormone feed preparation as described by Guerrero (1975) is done in three steps:

1 Hormone at a concentration of 30 mg/l is dissolved in 95% ethanol. Methyltestosterone is readily soluble in ethanol, but ethynyltestosterone requires several hours to dissolve completely at room temperature.

2 Finely ground trout feed is thoroughly mixed in an equal weight/volume ratio with the hormone solution.

3 The feed is then thoroughly dried. It can be dried in an oven at a temperature of up to 80°C. The feed then contains 30 mg hormone per kilogram of feed. Jensen (quoted by Shelton et al., 1978) suggests additives to enhance palatability, nutritive value, and stability. Antibiotics are also added to reduce losses of fish. These ingredients are added to the feed before drying. For 100 g of dry feed he adds 5 g cod liver oil, 0.04 g of uncoated ascorbic acid, and 0.1 g tetracycline. The feed should be refrigerated before use.

The fish are fed at 10–12% of body weight daily. The ration is divided into three to four feedings, or an automatic feeder operated by a timer is used.

After the completion of hormone treatment, the fry are transferred to an outdoor nursing pond (allowance should then be made for the expected loss since their weight is much lower than that of nontreated fry at the same age).

6.2.5 Nursing of Tilapia Fry

There are two ways to rear tilapia in ponds (for details see Section 7.3.3.): (1) rearing young tilapia in their first summer and (2) rearing nursed male fry in their second summer. In the first case, fry are transferred from the spawning pond into the rearing pond when they are over 1 g in weight. In the second case, the fry are nursed in ponds to a weight of 40–100 g, when it is easier to distinguish secondary sexual characteristics. Monosex hybrids are also nursed to 40–100 g in order to obtain fish with higher growth potential for the beginning of the next summer and to better utilize the natural productivity of the rearing pond (see Section 7.1).

Since the weight of fingerlings at the end of the nursing period will depend to a large extent on the length of nursing, those which are hatched and stocked in the nursery pond early in the season (end of May to beginning of June) can reach a final weight of 100 g and over. However, those hatched later and stocked in the nursery ponds in July–August will reach a weight of only 40–60 g. Stocking density is adjusted according to expected final weights. For example, when fingerlings are expected to reach a final weight of 100 g, the stocking density should be about 50,000/ha. In order to better use natural pond productivity, the fry are often stocked at a density of 100,000/ha for the first part of the nursing period (until the fingerlings reach a weight of about 50 g). Fry from the late spawns, which can reach a final weight of only 50 g, are also stocked at densities of 100,000/ha.

One of the important considerations in determining the desirable final weight of nursed fingerlings in regions with a cold winter is the available overwintering capacity. If specialized facilities and high investments are required for retaining adequate water temperature in the wintering ponds—such as covering the pond (Figure 6.7) or even warming the water—there is usually restricted wintering pond area and, therefore, a limitation on the standing crop of fish that can be held over the winter. Certain numbers of fingerlings are required for stocking rearing ponds in the spring. If we assume an average weight of 100 g for the fingerlings, multiply by the required number, and come up with an estimate that exceeds the standing crop that can be overwintered, then smaller fingerlings will be required (40–100 g). The stocking density of fry in the nursery ponds can be adjusted to achieve the final desired fingerling weight. These fingerlings can be nursed again in spring if necessary.

Losses of tilapia fry in the spawning and nursing pond are lower than those of carp. In their first stages of development the fry are

FIGURE 6.7. Covered pond for wintering of tilapia fingerlings.

protected by the mother, and until then no parasites or pests are found which cause serious losses of fry.

One of the problems encountered in spawning and nursing ponds is the presence of *Tilapia zillii*. This fish matures sexually at an early age (about 3 months) and small weight (10–15 g). It spawns a large number of eggs (up to 7000 per spawn). This fish not only competes with the fry of more desirable species, but it also is often transferred into rearing ponds with fish of the desired species since it is hard to distinguish between them. *Tilapia zillii* continue to breed in the rearing ponds and compete for the natural food, but they do not reach marketable size. This problem is especially severe on farms where the water supply involves a gravity flow system of open ditches. The small size of young *T. zillii* fry, after hatching, makes it hard to prevent their penetration into the pond through screens. This emphasizes, strongly, the importance of drying up all the pools in the spawning and rearing ponds before refilling.

6.3 CHINESE CARP

Three species of Chinese carp which have gained wide distribution— silver carp (*Hypophthalmichthys molitrix*), grass carp (*Ctenopharyngodon idella*), and big-head carp (*Aristichthys nobilis*)—are bred only by induced spawning. Since methods of breeding all three are more or less the same, a general description will be given here, following Pruginin and Cirlin (1973).

6.3.1 Selection of Brood Stock

No selection of brood stock is practiced as yet, apart from using healthy fish fitting the description of the species. Recently, however, large numbers of deformed fry have been noticed in Israel. This seems to be a result of inbreeding since the entire brood stock in the country came, in fact, from a limited number of fish. The best way to overcome this problem is to take females and males from different stocks.

For breeding itself, 3- to 4-year-old spawners, weighing about 5 kg each, are chosen. The females should have enlarged, soft bellies, and the males should have running milt. They should be chosen from stock held in ponds when the water temperature exceeds 26°C.

Eggs usually constitute about 18% of a ripe female's body weight. Each female contains about 700,000 eggs per kilogram of egg mass; thus a 5 kg female produces about 600,000 eggs. Only about 30% (i.e., 180,000) will hatch. Survival during the nursing period, both in the hatchery and in outside ponds, is about 60% in silver carp but only 20% in grass carp. Thus a spawning silver carp produces about 73,000 fry of 2 g each, and a grass carp produced only about 25,000 viable fry. Since only 7 out of 10 females respond to hormone injections, 20 females, 14 of which will spawn, are required to produce 1 million fry. The number of males required is about 1.5 for each spawning female.

6.3.2 Preparation of Pituitary Emulsion

The effect of the pituitary gland on gonad development is not species specific. Thus the pituitary of one species of fish can affect another species. The option of using fish pituitaries, which are easily obtainable, makes it unnecessary to use processed hormones. In Israel pituitaries are taken mostly from carp. The best time to prepare pituitaries is during the carp spawning season (April–May). Pituitaries are extracted only from females, since these are twice as potent as those from males. The fish selected are 700–1000 g. Extraction is done by killing the fish, cutting off the upper part of the skull, lifting the brain, and removing the pituitary with the help of forceps from the depression under the brain. The pituitary is washed in alcohol and then stored in pure ethanol until required. Drying the pituitary with acetone results in more rapid loss of potency. When required for injection, a number of pituitaries are emulsified in 1 ml of 0.75% saline.

6.3.3 Induced Spawning

Spawners are kept in separate tanks of about 0.7 m³ each. Before spawning the fish are treated prophylactically with formaldehyde against *Costia* infection. The fish are acclimated for 24 hours to hatchery conditions and then injected with pituitary emulsion. To minimize stress all injections are given while the fish are in the water. Fish are held in a scoop net in the corner of the tank and injected intermuscularly behind the dorsal fin. There are certain slight variations among hatcheries with regard to dosage and sequence of injections. The most common is an initial injection of 0.2 pituitary per kilogram of female body weight, followed by a second injection of 0.5 pituitary per kilogram 24 hours after the first. Eight hours later a third injection of 0.7 pituitary is given. Spawning occurs 8–12 hours after the third injection. A second version, which is also widely practiced, involves giving only two injections: the first is 0.3 pituitary, and the second, 24 hours later, is 1 pituitary per kilogram of female body weight. The males receive one injection of 0.6 pituitary at the time the first injection is administered to the females.

Incubation is carried out in incubation funnels as described above (see Section 6.1.6), but since the eggs of Chinese carp swell more than those of the common carp, only 200 g of fertilized eggs can be hatched in one funnel (i.e., about 140,000 eggs). Hatching occurs after 24 hours (at 24°C). After hatching the larvae tend to swim up and are usually swept by water flow to a receiving tank where initial indoor nursing is carried out for 10–11 days. After that the fry are transferred to a nursery pond.

Females are stripped while being held in a special restrainer consisting of a plastic 10 l container (40 × 20 × 20 cm). Milt is added and "dry mixed" with the eggs. The fertilized eggs are then washed in 3% saline (made with ground water) for 1 minute, after which the water is decanted. A second wash, of water only, is added, and the eggs are agitated for 2 minutes and again decanted. Three liters of water are then added, and the eggs are allowed to remain in the water for about 10 minutes without agitation. They are then poured through a large mesh net to remove any large pieces of tissue. The eggs, which have by now swollen considerably, are poured into a fine mesh net to remove excess sperm and washed. Finally, they are placed in incubators, where a steady flow of fresh water is provided to keep the eggs in motion but not to flush them out.

6.3.4 Nursing

Initial nursing is done in the hatchery in tanks similar to those used for holding the spawners. Upon yolk sac absorption, the fry are fed trout starter or soybean meal. After reaching 0.1–0.2 g they are transferred to earthen ponds for further nursing. Though they are less susceptible to attack by insects at this size, they may experience attack by *Notonecta* sp. Therefore, prophylactic treatment with Lindane (0.5 ppm) or oil (40–50 l/ha) is recommended.

6.4 MULLET

The lack of a reliable supply of fry is the major bottleneck in the culture of mullet in ponds. Dependence on natural spawning makes it difficult to plan pond stocking and fish production in advance. The study of induced spawning of mullet has reached quite an advanced state, and mullet fry have been produced in Taiwan, Hawaii, and Israel. However, full-scale commercial production has not developed. In most cases culture is still dependent on a supply of fry collected from natural habitats—estuaries and lagoons.

6.4.1 Induced Spawning

Induced spawning of mullet was studied in Israel during the 1960s by Yashouv and Berner-Samsonov (1970). They succeeded in spawning *Mugil capito* (*Liza ramada*), and a number of fry were produced. The techniques used are described by Pruginin and Cirlin (1973). Less success has been achieved with *M. cephalus,* where spawning has been induced but no viable fry have survived (Yashouv, 1969). Better success was achieved in spawning *M. cephalus* at the Oceanic Institute of Hawaii (Shehadeh et al., 1973; Kuo et al., 1974; Nash et al., 1974) and at the Tungkang Marine Laboratory in Taiwan (Liao, 1974). The reader is referred to these publications for details of procedures. In spite of the success achieved in spawning, the survival rate of fry was low. It seems that much further study will be required in order to improve the technique.

6.4.2 Collection of Fry from Natural Habitats

In most if not all places, mullet fry required for stocking ponds are still collected from estuaries during autumn and winter. In Taiwan, more than 10 million fry are collected from estuarine waters along the west coast during December through March. They are stocked in salt water, brackish water, or freshwater ponds (Liao, 1974).

Six species of mullet are found on the Mediterranean coast of Israel. Except for *M. labeo,* all species of fry enter the rivers and streams. There is a tendency for the mullet to seek different habitats within the rivers and streams. *Mugil saliens* fry are usually concentrated in the lower reaches of the rivers. *Mugil capito* and *M. cephalus* are concentrated initially near the mouths of rivers, but gradually disperse along the entire length of the rivers and are eventually found, in abundance, in the upper reaches. The size of *M. cephalus* fry, when appearing at the mouth of a river, is about 22 mm (0.3 g), and the size is about 18 mm (0.2 g) for *M. capito.* Only at this stage and size are they concentrated in schools where they can be caught in large numbers. When they become bigger, they are dispersed as individuals, or small schools, and it is much more difficult to catch them in quantity.

The most important factor to consider in facilitating fry collection of a more or less single species is the varied range of their seasonal appearances. Each species appears at the mouth of rivers at a certain season, and its run lasts for 3–4 months with a pronounced peak for a number of weeks. The sequence of appearances is regular, but the appearances and peaks may change somewhat in different years (Perlmutter et al., 1957; Pruginin et al., 1975). *Mugil saliens* appears in summer (July–September). Next comes *M. cephalus* in autumn and early winter (September–January), then *M. capito* in winter (January–April), and *M. auratus* in the spring (April–June). There is an overlap among these periods, so before the run of *M. saliens* ends, that of *M. cephalus* starts, and in samples of fry taken in August–September both species are found. In general, fish farmers start catching mullet for rearing in ponds when the percentage of *M. cephalus* is over 20%. *Mugil saliens* fry caught then and transferred to nursing ponds will not develop properly in fresh water and will disappear before *M. cephalus* are moved from nursing to rearing ponds. The conclusion of the *M. cephalus* run overlaps with the run of *M. capito,* and the end of the latter's run overlaps with that of *M. auratus.*

It is obvious that it is important to distinguish between species during the transition periods, especially in January where an overlap between the appearances of *M. cephalus* and *M. capito* may exist. Perlmutter et al. (1957) gave a useful key for the identification of mullet smaller than 10 cm. In spite of this key, it is difficult to rely on the external differences alone since these may change with the physiological condition of the fry. A much more reliable characteristic is the number of pyloric caeca on the stomach of the fry (see Figure 6.8). *Mugil cephalus* has two pyloric caeca, as compared to five to seven in the other species, which differ among themselves with respect to shape and order (see Perlmutter et al., 1957). In practice, the determination of species is achieved by taking a sample from the catch, dissecting each

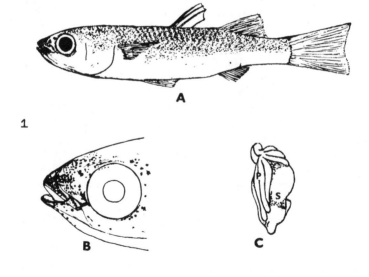

FIGURE 6.8.

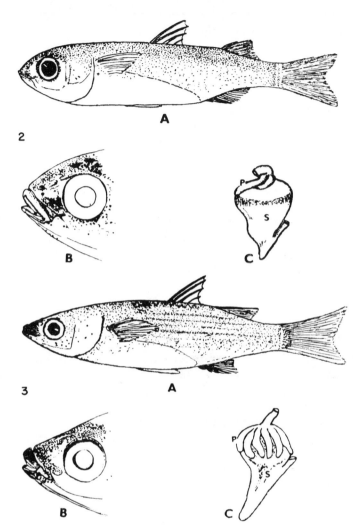

FIGURE 6.8. Morphological characteristics for the determination of three species of mullets at their fry stage: (1) *Mugil saliens,* (2) *Mugil cephalus,* and (3) *Mugil capito.* **A.** Side view of body: (1) Slender body and head; head less tapered than in *M. capito.* (2) Deep body. (3) Deep body, may be confused with *M. cephalus* but the head is more slender and tapered. **B.** Side view of snout: (1) Upper lip moderately thick; preorbital slightly bowed; teeth small. (2) Blunt snout; upper lip thin; a series of very small teeth present on premaxillary and mandible; inferior margin of preorbital slightly bowed. (3) Tapered snout; upper lip moderately thick; inferior margin of preorbital bone slightly bowed; teeth minute. **C.** Pyloric caeca (P): (1) In two groups of unequal sizes. (2) Two present, one large and one small. (3) Caeca equal in length, usually 6, rarely 5 or 7. (S = stomach.)

individual, and determining its species with the aid of a simple binocular microscope. The percentage of each species in the sample will determine whether or not the culturist should catch fish for stocking.

Mullet fry are concentrated in schools in the quiet waters of small indentations or lagoons along river banks where they rest before swimming upstream against the current. In these places they can be caught with a small seine of mosquito net mesh. During the peak of the run tens of thousands of fry can be caught each day. The number of fry caught is determined to a large extent by the weather. Major catches occur after sea storms and rainy periods. When there are only a few storms during the running period of any of the species (*M. cephalus* or *M. capito*), the number of fry entering the rivers is small. This is not corrected when stormy conditions develop later in the season.

Gray mullet are difficult to handle and transport. They should not be touched with the hands since this removes the protective mucous coat and scales and opens the way for various bacterial and fungal infections.

6.4.3 Nursing of Fry

From the adaptation tanks, or directly from the river, young mullet fry are introduced into relatively small ponds for preliminary nursing. Their density in these ponds should be about 30,000/ha. Medium-size male carp stocked at 200–300 fish/ha are also introduced into the pond to prevent development of filamentous algae. The mullet are nursed in these ponds for 60–100 days until the fry reach a weight of 1–3 g (usually by March–April). At this size they may be transferred directly to rearing ponds or to larger nursery ponds for further nursing.

As mentioned above, mullet are very delicate fish; they suffer and die quickly in muddy water, especially when water temperatures are 30°C or above (common around midday during summer). The transfer of young mullet fry from nursery ponds therefore must be done only in the early hours of the morning, when temperatures are still low. It is advisable to work quickly, with the help of a well-trained team, and have the proper equipment (nets, buckets, transport tanks, oxygen, etc.) ready for use before harvesting begins.

Since numbers of fry, especially those of *M. cephalus,* are limited, many fish farmers use *M. capito,* which is a slow grower and does not usually reach market size within 1 year. Therefore, there is a tendency to culture mullet into the second year of their lives. This is also done with *M. cephalus* because it reaches a larger size and is marketed at a higher price per kilogram (see Chapter 7).

In order to prepare fingerlings for the second year, fry from the preliminary nursing ponds are stocked for further nursing for the rest of the first year. *Mugil cephalus* will by then have reached a weight of 200 g, while *M. capito* will have reached a weight of only 20–100 g, depending on conditions in the pond (Pruginin et al. 1975). The density of stocking is about one-third to two-thirds that of initial nursing (10,000–20,000/ha). Since no feed is added to the pond, the carp growth ceases when their standing crop reaches about 80 kg/ha.

7   Culture Methods and Planning

The success of a fish farm is measured by its profitability. Profitability depends on the yield and market price of fish on the one hand, and on the cost of production on the other. These will, in turn, be affected by a number of factors, the most important of which are:

1   Stocking density of fish in the pond.

2   In polyculture—the species combinations and ratios.

3   Size of fish at stocking.

4   Size of fish at harvest.

5   Length of the culture period and time of stocking.

6   Fertilization and manuring.

7   Feed and feeding methods.

Most of these factors are interdependent. This complicates the determination of the optimal rate of each, and the balance among them, so as to maximize fish yield. For instance, when stocking density is low, natural food is usually sufficient to attain the maximum growth rate for the prevailing environmental conditions (section Section 7.1). Since the number of fish is relatively small, yield is also low. However, because little or no supplementary feed is required, feed cost is low. Increasing the density can result in increasing yields, but it often requires the supplementation of natural food with supplementary feeds, which are not always available and can be very costly. It is obvious that these relationships should be carefully thought out. Two groups of considerations have to be considered in such planning: (1) biological (especially ecological) and (2) economic.

7.1   BIOLOGICAL CONSIDERATIONS

Yield of fish per unit area is a function of growth rate and density. To increase yield one must know more about the factors involved and the interactions among them.

7.1.1 Growth Rate of the Fish

Two sets of factors affect the growth rate of individual fish:

1 Those related to the fish itself, such as its genetic characteristics and physiological state (state of health, sexual maturity, etc.):

2 Those related to the environment, of which the most important are:

    a. Chemical compositions of the water and bottom soil.

    b. Water temperature.

    c. Metabolite level (the products of excretion).

    d. Available oxygen.

    e. Available food.

Factors (a) and (b) are not affected by the presence of fish. In the majority of commercial fish farms they cannot effectively be changed by the fish farmer. Factors (c)–(e) are all either consumed or produced by the fish. Their concentration and thus their effect on the fish will be affected by fish density. When the supply of these factors per unit volume (or, in the case of the metabolites, the removal) is limited, then the higher the density of fish, the less food or oxygen is available per fish and the more metabolites are accumulated. In the case of food and oxygen, the fish farmer has a remedy—feeding or aeration of the water. The removal of the metabolites is more complicated, for it requires water flow through the pond.

As long as these density-dependent factors do not limit growth, the fish attain their maximum physiological growth potential for a given set of conditions (chemical composition of the water, water temperature, genetic characteristics, and physiological state). The larger the fish, the higher is their absolute potential growth capacity. However, this capacity is achieved only when enough food and oxygen are available and the metabolites do not inhibit growth. Figure 7.1 shows the absolute growth potential of carp as found from experiments at the Fish and Aquaculture Research Station at Dor, according to local environmental conditions (Hepher, 1975, 1978). Note that the absolute potential growth rate does not increase in a direct proportion to the increase in weight of the fish, but at a rather slower rate. This means that the relative growth rate, that is, the growth per unit weight (which can be expressed in percent), decreases with the increase in the weight of the fish. The following example relating to Figure 7.1 will demonstrate this. A fish of 500 g can growth much more (10.6 g/day) than a fish of 250 g (6.7 g/day). The latter, in turn, can grow

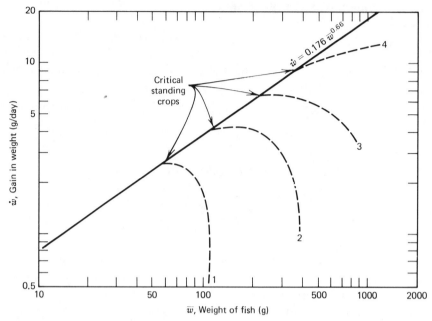

FIGURE 7.1. Average growth rate under different feeding and fertilization treatments at the Fish and Aquaculture Research Station, Dor, Israel. The heavy line represents potential physiological growth rate when food is not limiting. Broken lines note growth rate above the critical standing crops: (1) no fertilization or feeding; (2) chemical fertilization, no feeding; (3) chemical fertilization, feeding with sorghum; (4) chemical fertilization, feeding with protein-rich pellets.

more than a fish of 100 g (3.6 g/day). However, two fish of 250 g (= 500 g) will grow, under the same conditions, more than one fish of 500 g (6.7 × 2 = 13.4 g/day), and five fish of 100 g (= 500 g) will grow more than the larger fish (5 × 3.6= 18.0 g/day). If we express this growth as a percent of body weight, the 500, 250, and 100 g fish can grow 2.12, 2.68, and 3.6%, respectively.

The larger the fish, the more absolute amounts of food it requires in order to sustain its potential growth and maintain its body (Winberg, 1956). This is also true for the oxygen consumption and the production of metabolites. Since production of natural food is limited, when a pond is stocked with a certain density of fish they will reach a weight above which the natural food will not be sufficient to sustain both maintenance and the maximum potential growth rate. Because requirements for maintenance must be provided first, with the increase in fish weight over this point the growth rate will be less than the potential maximum, and when the food deficit increases, growth rate will be reduced. This is the *critical standing crop* (CSC) of the fish (Hepher, 1975, 1978). Growth will cease entirely at the *carrying capacity* when available food is just sufficient to maintain the fish but insufficient to allow growth. If at this point the fish are below market size the total yields cannot be marketed as food fish. In order to maintain the potential individual growth rate beyond the CSC, the amount of food should be increased either by fertilization or by adding supplementary feed. Another way to maintain the individual growth rate is to reduce the density of fish so that the available food will be divided among fewer fish. At the CSC point it may be only one of the components of the food which becomes exhausted and turns into a limiting factor that inhibits further growth. Other components of the food may still be present in sufficient quantities. Adding rather simple supplementary feed which provides the missing component can restore growth to its potential until another component becomes a limiting factor. For instance, natural food is very rich in protein (50–60% of its dry weight). When cultured on natural food only, carp (and this is true also for many other fish) consume more protein than they require. However, at a certain standing crop, energy becomes limited and the fish utilize the protein as an energy source. When carbohydrate-rich feed, such as cereal grain, is offered, natural protein is released for growth. Thus, by feeding simple, low-cost feeds one gets in return high-quality protein as fish flesh. This simple feed can affect growth as long as the natural food proteins are sufficiently available for growth. When fish weight increases to a standing crop over which the demand for protein cannot be met by natural food, growth will again be limited unless more supplementary protein is added. The same is true for vitamins and minerals. As long as they are provided in sufficient quantity by natural food, there is no need to include them in the supplemental diet. However, if deficits are developed, growth will decrease unless they are added. It is obvious that with an increase in standing crop, both the quantity and the quality of supplementary feed must be altered (see Chapter 10). For every feed there is a different *critical standing crop* (CSC), above which growth is inhibited, and a different carrying capacity at which growth ceases (Figure 7.1).

Food is not the only factor that affects CSC and carrying capacity. As standing crop increases, the supply of oxygen may become insufficient, especially during the early morning, and may influence CSC (see Section 11.1). This will also happen as metabolites accumulate. However, inhibition of growth by these factors occurs at a much higher standing crop than it does with the lack of natural food.

From experiments at Dor, critical standing crops and carrying capacities have been estimated for common carp at different feeding levels. See Table 7.1. Much higher carrying capacities have been given by Van der Lingen (1959) for *Sarotherodon mossambicus* in Rhodesia:

Unfertilized ponds	896 kg/ha
Fertilized ponds	2128 kg/ha
Fertilized and fed supplementary feed	6160 kg/ha

The differences in carrying capacities seem to indicate that the amount of natural food available for tilapia is much higher than that available for common carp, and therefore tilapia will require added supplementary feed only at much higher standing crops than the carp (see Section 10.3).

TABLE 7.1   Effect of Feeding Level on Critical Standing Crop and Carrying Capacity at the Fish and Aquaculture Research Station, Dor, Israel (Hepher, 1975)

Feeding Level	CSC (kg/ha)	Carrying Capacity (kg/ha)
No feeding and no fertilization	65	130
Fertilization but no feeding	120	480
Fertilization and feeding with cereal grains	550	2000
Fertilization and feeding protein-rich pellets	2400	—

7.1.2 Density

Both CSC and carrying capacity are expressed as biomass in kilograms per hectare. Since biomass is a product of the average individual weight of the fish and their number per unit area, increasing each of these factors increases food consumption of the population. From experience in ponds it seems that the CSC and carrying capacity values will be more or less unchanged whether the pond population is composed of a few large or many small fish (for discussion on reasons see Hepher, 1978). The following example is based on the CSC and carrying capacity given above for carp in fertilized ponds fed cereal grains. If such a pond is stocked with 2000 carp/ha, CSC will be reached when the fish attain an average weight of 275 g (0.275 × 2000 = 550 kg/ha). Below this weight the potential growth rate has been achieved, but above 275 g their growth rate will be less than the possible potential due to lack of food. When the fish attain 1 kg (= 2000 kg/ha) they will cease growing. If, however, the same pond is stocked with 4000 carp/ha, they will reach CSC at 137 g and will cease growing at 500 g.

Since below CSC the growth rate of fish of a given weight reaches a maximum, the yield is proportional to density. The higher the density the higher the yield (Figure 7.2). When CSC is reached the growth rate decreases and the log yield–log density function deviates from a linear proportion. As long as growth rate decreases at a smaller rate than the increase in density, yield increases. Above a certain standing crop the average growth rate decreases at a higher rate than the increase in density and the yield decreases sharply and reaches zero at carrying capacity. Maximum instantaneous yield (for a given weight) is reached at a standing crop *between* the CSC and the carrying capacity. However, as the fish continue to grow and standing crop is constantly increasing, carrying capacity will soon be reached and growth will cease. That means that allowance should be made for further growth

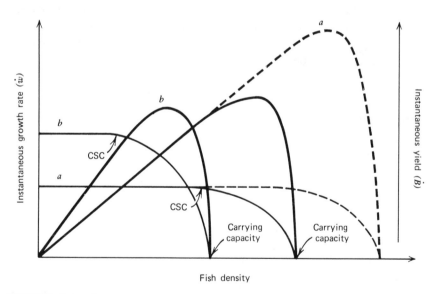

FIGURE 7.2. Schematic representation of the relationships among fish density, instantaneous average growth rate ($\dot{w}$), and instantaneous yield ($\dot{B}$) of fish of two different average weights ($a < b$). Heavy lines represent yields. Broken lines show supplementary feed added.

by stocking at a lower standing crop even if this means a lesser utilization of the natural productivity of the pond. Such lowering of standing crop can be achieved only by adjusting the density to the required final average fish weight at harvest.

The following example will illustrate this. When carp are cultured commercially in Israel only on cereal grains, a maximum standing crop of 1200–1500 kg/ha is usually attained (carrying capacity is not reached). If the fish are to be marketed at an average weight of 600 g (= 0.6 kg), and no intermediate harvesting is planned, the stocking rate will be 1500/0.6 = 2500 fish/ha.

The density can be used for regulating the average growth rate of the fish and, therefore, the length of the rearing period. From the foregoing discussion it is clear that for a given feeding regime, the size of the fish alone will affect growth rate below the critical standing crop. The larger the fish, the more it will grow and stocking rate will make no difference. However, when stocking rate is increased, CSC will be reached at a lower fish weight and the growth rate above CSC will be reduced. The average growth rate for the entire rearing period will thus be lower.

The most economical stocking rate is not necessarily that which results in the highest average growth rate, but rather that which results in the highest yield per unit area. This can best be explained by the following example. Assume that the average growth rate of fish for the growing period in a pond stocked with 200 carp/ha is 6 g/fish/day. The yield will then be

$$2000 \times 6 = 12 \ \text{kg/ha/day}$$

If by increasing the density to 3000 carp/ha the average growth rate is reduced to 5 g/fish/day, then the yield will be

$$3000 \times 5 = 15 \ \text{kg/ha/day}$$

In such a case, it is more economical to stock 3000 carp/ha in spite of the reduced growth rate.

A word of caution should be given here. Overstocking may lead to two pitfalls:

1 Carrying capacity may be reached and growth will cease entirely before the fish have attained market size.

2 Beyond a certain stocking density any increase in stocking rate will cause a proportionately larger reduction in growth rate.

Thus, if the density in the previous example was increased to 4000 carp/ha and the average growth rate was reduced to 2.5 g/fish/day, the yield would only be

$$4000 \times 2.5 = 10 \;\; \text{kg/ha/day}$$

As mentioned above (Section 7.1.1), the growth rate of fish is not constant. It increases with fish weight. However, for practical purposes, and for a given range of growth, it can be expressed as an average growth rate for this range in grams per day. This helps the farmer to compare growth rates of fish in different ponds when the growing period in these ponds is not equal, or to compare the actual growth rate obtained in a pond with the expected growth. Table 7.2

Table 7.2 Growth Rates of Various Fish Species at Specific Densities and Weight Ranges in Israeli Fish Farms

Species	Density (fish/ha)	Range of Fish Weight (g)	Approx. Average Growth Rate (g/day)
Common carp	4000–6000	100–650	5
Tilapia	3000	5–200	1.5–2
Tilapia (males)	2000	100–500	3
Silver carp	500–1500	200–1500	7–10
Mullet	1000	2–500	1.5

summarizes the expected growth rates of various species under Israeli conditions for specified ranges and densities, when fish are fed protein-rich pellets.

Density will depend, on the one hand, on the market size of the fish: the smaller the market size, the higher the stocking density. On the other hand, density will be strongly dependent on pond productivity—the amount, quality, and level of management (aeration, water flow, etc.). Walter (1934), the founder of modern theories on pond management and production, divided ponds into three categories, according to their productivity. He suggested that ponds in the lowest productivity category be used for nursing rather than rearing since final weight of the fry is less crucial than the final weight of market fish.

The higher the productivity or level of supplementary feeding, the higher the density that can be maintained in rearing ponds. It follows that maximum obtainable yield will be correlated to fish density which, in turn, will be affected by feeding level (and the natural productivity of the pond). This means that when food is available,

increasing density will result in increased yield. Figure 7.3 presents
the relationship between stocking rate and maximum attainable yield
in ponds stocked with carp or in polyculture where carp is the main fish
and appropriate management is provided.

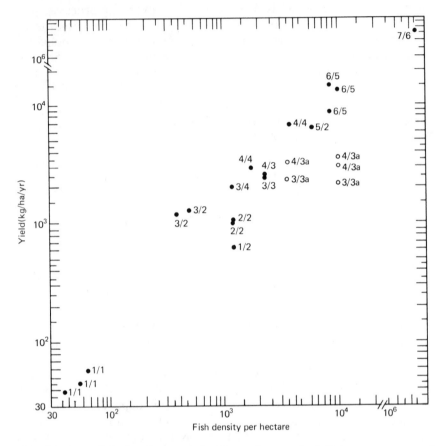

FIGURE 7.3. Relation between density and annual yield in some carp and
polyculture ponds. Numerator of designating fractions represents the treatment: 1, no feeding or fertilization; 2, chemical fertilization only; 3, fertilization and feeding with cereals; 4, feeding with a protein-rich diet; 5, feeding
with a protein-rich diet through a demand feeder, 6, feeding with protein-rich
diet, aerated ponds; 7, flowing water system. Denominator of designating
fractions represents source: 1, Nakamura et al., 1954; 2, Hepher, 1962; 3,
Szumiec, 1979; 4, Yashouv, 1969; 5, S. Sarig and Marek, 1974; 6, Kawamoto,
1957. The interpolated line formed by the black dots represents the maximum
yield obtained when no environmental and food conditions limit the growth of
fish (for open circles see text).

When density is low, corresponding to a pond's natural productivity, fertilization, manuring, or feeding without correspondingly increasing stocking density will result in only a limited increase in yield, if any. If represented on the graph in Figure 7.3, the point giving the yield of such a pond would probably stay on the interpolated line, but at a lower yield than could be expected according to the treatment applied. If this is the case, it shows that the extra food provided by the treatment was not used, and a higher yield could have been obtained by increasing the stocking rate. On the other hand, increasing the stocking density without a concurrent increase in food resources will result in overutilization of the food and reduced growth. If the carrying capacity is reached, growth will cease. This is represented in Figure 7.3 by the points 3/3a and 4/3a, taken from Szumiec (1979). These ponds received the same treatment as 3/3 and 4/3, but in spite of the higher stocking density they did not produce higher yield.

Polyculture is a way to utilize better natural productivity with a concurrent increase in density with fishes of different species. Yashouv (1971) introduced 1000–2000 tilapia (S. aureus) per hectare in addition to the common carp population. As a result the yield increased considerably without affecting the growth of carp. Further increases in stocking density by adding silver carp and other fishes result in even higher yields. In Israel, polyculture systems include 3000–6000 carp, 3000–4000 tilapia, 500–1500 silver carp, about 1000 mullet, and about 300 grass carp/ha, so that the total stocking density reaches from 7000 to 9500 fish/ha. (for further discussion on polyculture see Section 7.3.1).

Aeration makes it possible to further increase stocking density. Two aeration systems have been developed. In *intensive polyculture,* where stocking density is about 10,000–12,000 fish/ha, auxiliary aeration is used as necessary and only during the critical morning hours. The second method is that of the *highly intensive* pond culture, where stocking density reaches 15,000–20,000 fish/ha, either in monoculture of carp or tilapia, or polyculture of these two species. Because of the high densities, aeration is obligatory at least during the greater part of the night and the early morning. Yields up to 20 ton/ha/yr can be obtained by the latter method (Marek and Sarig, 1971).

The correlation line shown in Figure 7.3 is specific to the single species or combination of species in a pond. It can differ with environmental conditions and fish market size. Since the stocking rate in most ponds is not properly adjusted to the food resources, most of the points describing a fish density to yield relationship will fall below the line. The line will represent, therefore, the maximum yield attainable at a given treatment and stocking rate for a given fish combination and environmental conditions.

From the foregoing discussion the following conclusions may be drawn:

1 Increasing the yield per unit area (and unit volume of water) is possible only by intensification, that is, by increasing stocking density on the one hand and the inputs (fertilization, feeding, improved diets) per unit yield on the other. These two factors must, however, be dealt with simultaneously.

2 Since larger fish have higher absolute potential growth rates at a given stocking density, higher yields can be obtained during a given period of time by stocking bigger fish, provided that they have sufficient food (and oxygen). However, when a standing crop is limited by food, higher yields can be obtained by stocking a higher density of smaller fish rather than a lower density of larger ones, since the relative growth rate (i.e., growth rate per unit of weight) is greater with smaller fish. This can be illustrated by the following example. When carp are fed cereal grains and market weight is 600 g, the density is

$$\frac{1500}{0.6} = 2500 \text{ fish/ha.}$$

The critical standing crop is reached at about 550 kg/ha, or with 2500 fish/ha at about 220 g/fish. Up to this weight the fish are expected to grow at maximum rate. A comparison of the potential daily yield per hectare when the average weight of fish is 10 g with that obtained when their weight is 100 g will show:

a. At 10 g the maximum potential growth rate according to Figure 7.1 is 0.8 g/day, and the yield per hectare will be
$$0.8 \times 2500 = 2000 \text{ g/ha/day}$$

b. At 100 g, the maximum potential growth rate per fish is 3.6 g/day
$$3.6 \times 2500 = 9000 \text{ g/ha/day}$$

It can be seen that at a given density, when enough food is available, larger fish will produce a higher yield. Thus, in the above example, it will be more productive to stock 100 g fish and rear them to 600 g than to start with 10 g fish. No doubt stocking very small fish will be even less productive. This will change when the standing crop increases over the CSC and the growth rate per fish, as well as yield per hectare get lower as the carrying capacity is approached (Figure 7.4).

However, when the market accepts fish of 300 g (= 0.3 kg) each (as is the case of fish marketed for processing), the stocking density can be increased accordingly:

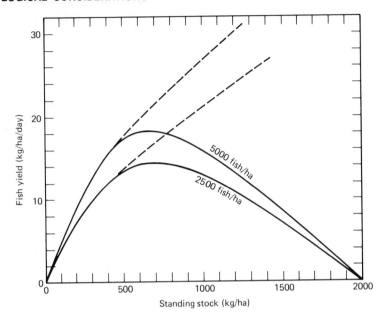

FIGURE 7.4. Average instantaneous yields (kg/ha/day) of carp fed cereal grains at two densities (2500 and 5000 fish/ha) when the standing crops are equal. Growth rates are from experimental results at the Fish and Aquaculture Station, Dor (see Figure 7.1). Broken lines represent possible yield when food does not limit growth.

$$\frac{1500}{0.3} = 5000 \ \text{fish/ha}$$

The daily yield of fish of 100 g at a growth rate of 3.6 g/day (Figure 7.1) will then be

$$3.6 \times 5000 = 18,000 \ \text{kg/ha/day}$$

and that of 200 g at 6.7 g/day will be

$$6.7 \times 5000 = 33,500 \ \text{kg/ha/day}$$

The yield for the entire growing season will accordingly be higher (Figure 7.4). This is another justification for nursing fry at higher stocking densities. The production per unit area in nursing ponds is much higher than in rearing ponds, and the area is utilized more efficiently.

3 The closer the standing crop gets to the CSC, the better the utilization of natural food in the pond. This again means that at a given density the larger the fish, approaching the CSC, the higher the gains obtained with relatively less supplementary feed (i.e., a lower feed conversion ratio). This is a further justification for nursing fry before the rearing period.

Increasing the weight of stocked fish as implied by conclusions 2 and 3 above means a short rearing period. It also involves frequent draining of the pond, an increase in costly harvesting operations, loss of growth time used for pond refilling, and much more handling of the fish. A compromise should be found, therefore, between the two. The usual practice is to stock fish of about 100 g and rear them during a growing season of about 120 days.

4 Another way to use natural pond productivity is by stocking at a high density and thinning the population when carrying capacity is approached. This also results in higher yield. However, frequent thinning, besides the extra work involved, means that the fish which were taken out are still under market size; they are transferred to another pond for further growth and will then have a short growing season with the extra costs involved, as discussed above.

From the above discussion it is clear that fish density and its control are of prime importance in pond management. To sum up, the factors involved are listed below. The interaction among these factors will determine the desirable stocking density:

1 Weight of marketed fish.

2 Natural pond productivity.

3 Feed and feeding level—whether simple feeds such as cereal grains or compounded protein-rich pellets.

4 Method of culture—monoculture or polyculture.

5 Length of the rearing period and whether or not intermediate harvesting is planned.

The density of fish in a pond can change spontaneously during the growing period without the knowledge of the fish farmer. Diseases or anoxia may drastically reduce the number of fish, while uncontrolled "wild" spawning may increase it. Both may lead to decreases in yield and wasted feed.

Mortalities are often correlated to the size of fish, since smaller fish are more susceptible to parasites, diseases, and rough handling. Most of the losses occur, therefore, immediately after stocking or during the early part of the rearing period.

"Wild" spawning is common in many ponds and may reduce the yield to uneconomical levels. Due to the competition by the great number of small fry, the whole population becomes stunted and growth ceases. "Wild" spawning usually occurs in the later part of the growing season when the fish reach sexual maturity. It is more of a problem where the fish mature early, such as occurs with tilapia in warm climates, or where the fish are sold at larger sizes. In any case, every effort should be made to avoid both mortality and "wild" spawning.

7.1.3 Length of the Rearing Period

The length of the rearing period is sometimes a constraint imposed on the fish farmer. If the fishpond also serves as an irrigation reservoir, irrigation requirements will usually affect the length of time in which pond water can be held. There are usually two kinds of these reservoirs: those which are completely drained at the peak of the irrigation season (such as in cotton irrigation in Israel, where the reservoirs are drained near the end of July or beginning of August); and those which are held until the new rainy season, in November. In the first case the length of the rearing period is more or less normal for fish culture—120 days. In the latter case, the fish have a longer period for growth and will attain higher individual weights. Thus, fish stocking density should be higher at first to reduce individual growth, but the population should be thinned a number of times during the rearing period so that the carrying capacity is not reached.

In the majority of cases, length of the rearing period is determined (within a certain range) by the fish farmer himself. Factors affecting this include:

Minimum Weight of the Marketable Fish.   This may change with time through a change in consumption habits, and it differs by region, but it is usually quite constant within regions. In Israel today the following minimum fish sizes are accepted by the market: carp, 600 g; tilapia, 200 g (larger fish get higher prices per kilogram); silver carp, 1500 g; and mullet, 500 g. In Germany the market will accept mainly larger fish—carp of 1000–1500 g—while in many of the Southeast Asian countries a tilapia of 100–150 g or less is readily acceptable in the markets. It is obvious that at a given growth rate, a longer rearing period is required to produce larger fish. The only way to change the length of the rearing period is to vary the size of fish stocked. This will, of course, affect the length of the nursing period. Alternately, stocking density may be altered so as to affect the growth rate of the individual fish.

Average Growth Rate of Individual Fish. This will be affected by the size of fish reared, stocking density, and feeding regime (see Section 7.1.2). If the average growth rate of the fish is 6 g/day, starting with a fish of 100 g with a target of 600 g, it will take 83 days (500/6) to attain market size. If, however, the growth rate is reduced to 5 g/fish/day, it will take 100 days (500/5) to reach the same target. When the rearing period is limited, stocking rates must be reduced to increase individual growth rates, even if this may also mean lower yields per unit area.

By considering these two aspects, we can calculate the minimum length of the rearing period required to produce market-sized fish. For conditions in Israel this period is about 120 days, which means that the fish can be reared for two such periods during a summer. In many cases, however, the rearing period is longer than that. One reason is the need to distribute harvesting throughout the summer, both because of limitations in manpower and to avoid the costly storage of live fish in special ponds before marketing. In order to achieve this aim, some ponds must have a longer rearing period. In other cases the pond is utilized for nursing or storing fish for a period shorter than 120 days. The rearing period is then extended so as to close the time gap. It should be remembered that the "target" standing crop at the end of both the short and the long rearing periods is more or less the same, and the only differences are higher fish weight and lower density at the end of the longer rearing periods as compared to the shorter ones. The marginal yield per unit area is lower when the rearing period is longer. In spite of this loss of potential yield, lengthening of the rearing period is necessary and may even be beneficial from the point of view of general planning and the economics of operating the farm.

7.2 ECONOMIC CONSIDERATIONS

Four main economic factors, besides the possession of sufficient investment capital, affect the management methods chosen and the profitability of the farm. These are the availability and costs of land, water, labor, and supplementary feed and fertilizers. In Israel, both land and water are in rather short supply. The main shortage is in the allocation of water and its supply to agriculture. There is, therefore, constant competition for water, and products that bring the highest income per unit of water win. Fishponds that are generally part of a much larger integrated agricultural operation must also compete for water.

Another economic factor affecting the culture method is the nature of fish production costs. One should distinguish here between costs related to the farm area and those related to fish yield (see Chapter 11). Costs of initial construction of ponds and water system, equipment, water, and a certain amount of labor are related to farm size, while costs for feed, fry, harvesting, and marketing are related to fish yield. One characteristic of any fish farm is high capital investment in pond construction, water systems, and equipment.

Both the need to compete for water and the need to cover area-related costs demand that high yields per unit area be obtained. Calculating the break-even point of the area-related costs, in terms of equivalent fish yield (after deducting the costs for fry, harvesting, and marketing from the income), shows that in Israel yields of 1740 kg/ha of carp or 2440 kg/ha of tilapia are required to cover these costs. This yield cannot be reached on natural food only (Tal and Hepher, 1967), and intensification is required in order to increase the yield per unit area and per unit of water. This, in turn, calls for an increase in stocking density, on the one hand, and an increase in inputs of feed, fertilizers, manures, aeration, and other factors, on the other. However, as stocking density increases less natural food is available per fish, resulting in the need to increase levels of supplementary feed. The quality of this feed must also improve with density, until at high density, a nutritionally balanced feed, containing all necessary vitamins, minerals, and nutrients, must be provided. Such feed is used for the highly intensive carp culture at Tanaka farm in Japan (Tanaka, personal communication; Chiba, 1965) and in trout culture. At very high densities the accumulation of metabolites may play a large role in growth inhibition (unless removed by a large flow of water, as at Tanaka farm and trout farms). In addition, the feed conversion ratio is high. The market prices of the warm-water fish in most places do not justify these costs. This means that some intermediate intensification level should be employed in which adding manure or feed increases the yield per unit area and lowers the area-related costs per ton of fish yield, but does not increase the yield-related costs beyond an economical level. No doubt polyculture, which includes fish that utilize natural food to a greater extent and do not consume supplemental feed (fish such as silver carp, bighead carp, and to some extent tilapia), helps to achieve this.

7.3 MANAGEMENT METHODS

Production targets, environmental conditions, and socioeconomic situations determine the method of management. If the goal is production of low-cost fish which can be afforded by a wide part of the population, inputs must be low. The stress then is on greater utilization of natural food. This in turn means stimulating the production of natural food by chemical fertilizers or organic manure, on the one hand, and better utilization of natural food through polyculture, on the other. Nevertheless, fish density and thus yield per unit area must be limited.

If, however, the goal is production of a more expensive fish, which can pay for higher inputs such as high-quality feed, running water, and others, then fish density can be considerably increased with a parallel increase in yield per unit area. Natural food plays a much smaller role, if any, and fertilization and manuring become less effective. Polyculture has no advantage over monoculture. On the contrary, monoculture of a single species and a more or less uniform-size fish saves sorting and grading and makes the culture simpler.

7.3.1 Polyculture

The first and most important consideration in polyculture is the probability of increasing production by better utilization of natural food. Choice of species and their densities will be determined accordingly. In order to achieve increased production, the species stocked must have different feeding habits and occupy different trophic niches in the pond. Common carp, tilapia, silver carp, grass carp, and, to a certain extent, mullet and bighead carp all differ in their feeding habits, and their concurrent culture in a pond increases total fish yield.

In addition to the above considerations, other factors—some positive and others negative—should be taken into account. Yashouv (1971) demonstrated that the yield of silver and common carp, cultured together in polyculture, is higher than that of either species alone. A similar effect has been seen when culturing common carp and tilapia. The following explanations have been given for these synergistic effects:

1 Improvement in pond oxygen regime occurs due to the presence of silver carp and tilapia. Silver carp consume excess algae. When the concentration of algae is excessive, an imbalance between production and consumption of oxygen is created (see Section 11.1.2). The consumption of excess algae by silver carp improves and balances this situation. It does not reduce oxygen production to any great extent but does considerably decrease oxygen consumption (Barthelmes, 1975). Tilapia feed on the organic ooze of the pond bottom. This ooze increases oxygen consumption when it is decomposed by bacteria. Thus tilapia also improve oxygen conditions in the pond.

2 Some fish feed on the excreta of other fish. This has still to be studied, but it seems that tilapia feed on the excreta of common carp which feed on excreta of silver carp. The latter do not digest all the algae that they consume. The larger excreta "pellets" make these algae available to common carp which cannot consume free algae in the water.

3 Another positive effect of polyculture is the decrease in the population of *Tilapia zillii* due to the presence of the mouth brooder *Sarotherodon aureus*. There is distinct competition between these two species in which the latter depresses the population of the former.

In contrast to these positive effects, there are some negative effects of polyculture which also should be pointed out. These usually consist of competition among the different species when an imbalance is created. From an economic point of view one should recognize that sorting the different species during harvesting is an expensive operation requiring special equipment (see section 8.43) and considerable manpower. If the extra income received from higher overall yields does not cover these expenses, the establishment of polyculture is not worthwhile.

Obviously the ratios among the different species and their stocking densities are or prime importance in polyculture. These can be determined in two ways: (1) If one knows the food habits of the fish in the polyculture system, and the amount of natural food available in the pond, then the relations among the species can be calculated. This was done by Tang (1970) in Taiwan. (2) A second way is by trial and error, which is the way most commonly used for determing these relationships.

In practice one should take into account not only production but also the market price of each of the fish produced and the potential income from the different possible combinations, as well as management considerations, such as the possibility of intermediate harvesting and the size differences among species. The growth rate of each species should be such that all of the fish will be harvested simultaneously at market size and there will be no need to transfer fish for further growth, since such transfer is sometimes associated with a large loss. Rate of growth will depend on stocking density and will, therefore, affect the proportions among species.

From the foregoing discussion it is clear that different combinations of species in polyculture may be used in different regions according to specific conditions and preferences of the market, and such differences can also be found between farms or, to some extent, even on the same farm. It is not surprising, therefore, that Hoffman (1934) gives 11 such combinations which were used by Chinese farmers before the Communist revolution. Lin (1949) mentions five combinations for South China and three others for Hong Kong. Hora and Pillay (1962) give a number of other combinations for various districts in China and neighboring countries. Tapiador et al. (1977) give the following polyculture combinations practiced in the Pearl River Delta of China:

1 Where the major species is grass carp:

Grass carp (*Ctenopharyngodon idella*)	55%
Silver carp (*Hypophthalmichthys molitrix*)	16
Bighead carp (*Aristichthys nobilis*)	10
Others	19

2 Black carp as the major species:

Black carp (*Mylopharyngodon piceus*)	42%
Grass carp	24.2
Silver carp	12.4
Bighead carp	7.4
Wuchan fish (*Megalobrama amblycephala*)	7.4
Common carp (*Cyprinus carpio*)	3.4
Golden carp	3.2

3 Silver carp as the main species:

Silver carp	65%
Bighead carp	10
Grass carp	12
Common Carp	5.2
Wuchan Fish	7.8

Total stocking density in these ponds was about 15,000/ha.

Some combinations of polyculture that gave high yields in Israel are presented in Table 7.3.

7.3.2 Monoculture

It is clear that at relatively low densities monoculture does not fully utilize natural pond productivity. However, when stocking density is very high, the relative role of pond productivity in the overall nutrition of fish decreases since existing natural food resources have to be divided among more individuals. The gain which can be obtained by polyculture is relatively small, while the extra work involved with sorting the different species of fish at harvest time becomes a burden. Monoculture is therefore the only method of culture used in running water systems and in cages where the supply of natural food is limited. In ponds high stocking densities are not common. Oxygen is usually one of the limiting factors for increasing fish density. However, aeration of pond water makes it possible to overcome this limitation and to stock ponds with high densities. Such a method has been developed recently in Israel. Stocking rates are relatively high, reaching 20,000 fish/ha and above. This method of culture is still under development and has not yet spread widely. Monoculture, either of carp or more often of tilapia, is practiced in aerated ponds. Some difficulties usually arise in such ponds. The following are a few of the major drawbacks:

Relatively Low Growth Rates of Small Carp. It seems that smaller fish are dependent to a large extent on natural food and do not respond as well to supplemental feed as do larger carp (Hepher, 1978). High stocking density results in much smaller amounts of natural food per fish. This cannot be compensated for by supplementary feed unless it is a complete diet.

Short Rearing Period. The rearing period is often very short since a drastic drop in growth rate is noticed after about 60–100 days. The reason for this is not clear. It may be the result of the accumulation of metabolites which seem to be higher in ponds stocked at high densities. It may also result from the accumulation of reduced materials in the pond bottom resulting in a lowering of the redox potential of the soil and the accumulation of methane and hydrogen sulfide (H_2S) which are poisonous to both fish and natural food organisms.

The accumulation of reduced material in the bottom seems to affect carp, which are bottom feeders, more severely than tilapia. The growing period of the former is shortened to about 50 days, while the tilapia can grow for a longer period—about 100 days. In addition, when monoculture of carp is attempted, the demand for oxygen at the end of the growing period cannot always be satisfied by aeration, and there is a severe danger of anoxia.

Increase in the Feed Conversion Ratio. Since the amount of natural food per fish is rather low, if the feed used to supplement it does not include all the necessary nutrients for growth, growth rate will be reduced. Reduction of growth due to nutritional insufficiency and accumulated metabolites results in increased feed conversion ratio, which may reach an uneconomical level.

Filamentous Algae. A problem which occurs in tilapia ponds is the development of filamentous algae, which becomes a nuisance when the fish are harvested. In order to overcome these difficulties, a number of common carp are introduced into tilapia ponds (up to about 20%).

Similar to the introduction of common carp into the tilapia "monocul-

TABLE 7.3 Polyculture Combinations Resulting in High Yields in Some Israeli Fishponds

Species	Stocking Date	Harvesting Date	Rearing[a] Period (days)	Density at Harvest (fish/ha)	Average Stocking wt. (g)	Average Harvesting wt. (g)	Yield (kg/ha)
Farm: *Gan-Shmuel*							
Pond area: *4.2 ha*							
Two rearing periods							
Common carp I	Feb. 24	Aug. 6	143	3,000	50	942	2,670
Common carp II	Apr. 6	Aug. 6	122	3,444	15	230	730
Tilapia	Apr. 5	Aug. 6	123	2,020	148	487	670
Silver carp	Feb. 24	Aug. 6	143	1,720	200	1311	1,910
Total of first rearing period:				10,184			5,990
Common carp I	Aug. 8	Dec. 12	99	3,230	387	1018	2,040
Common carp II	Aug. 11	Dec. 12	96	3,120	10	235	700
Tilapia	Aug. 8	Dec. 12	99	2,850	48	288	680
Mullet	Aug. 22	Dec. 12	85	1,180	160	430	350
Silver carp	Aug. 22	Dec. 12	85	1,600	416	1000	930
Total of second rearing period:				11,980			4,670
Annual yield (242 days):							10,660

Farm: Hama'apil
Pond area: 5.7 ha
One long rearing period
with intermediate culling
of carp and nursing of
tilapia toward the end of
the period

Common carp	Mar. 12	Dec. 29	245	4,540	29	1106	4,889
Mullet	Mar. 17	Dec. 29	244	1,750	156	500	563
Silver carp	Mar. 7	Dec. 29	245	1,400	195	2268	2,980
Tilapia	Aug. 9	Dec. 29	98	9,120	13	100	835
Annual yield (245 days):							9,267

Farm: Lehavot Habashan
Pond area: 5.2 ha
Two rearing periods, the
first combined with
nursing of tilapia

Common carp	Jan. 1	Aug. 30	168	3,850	100	953	3,240
Tilapia	May 5	Aug. 30	118	11,440	28	170	1,620
Silver carp	May 5	Aug. 30	118	670	300	1650	1,280
Total of first rearing period:				15,960			6,140
Common carp	Sept. 9	Oct. 28	50	2,500	210	430	440
Tilapia	Sept. 1	Oct. 28	50	3,300	170	280	730
Silver carp	Sept. 9	Oct. 28	50	420	700	1400	210
Total of second rearing period:				6,220			1,380
Annual yield (218 days):							7,520

[a] Calculated length of rearing period does not include days prior to Mar. 15 and after Nov. 15 due to cold weather.

ture" ponds, it is also common to introduce a number of tilapia into the common carp ponds. The tilapia consume part of the accumulated organic matter and thus ease the oxygen demand. The monoculture system thus becomes a "bi-culture" system, in which one species is dominant.

The growth rate of tilapia is affected by stocking density to a lesser extent than that of carp. While tilapia at densities of 3000–4000 fish/ha in polyculture can gain an average of up to 5–6 g/day (at a weight range of 80–400 g and when fed with pellets), daily gain drops when density is increased, and it levels off at a density of 20,000 fish/ha to about 2 g/fish/day. In spite of high density, the feed conversion ratio remains relatively low, while yield is high. This explains why tilapia "monoculture" is more common in Israel than that of common carp. Examples of both tilapia and carp "monocultures" are given in Table 7.4.

7.3.3 Culture of Tilapia

Tilapia, though valuable pond fishes, pose a special problem when reared in ponds, whether in polyculture or monoculture. The overprolif-eration and "wild" spawning in rearing ponds result in such large numbers of small fry that stunting of the entire tilapia population occurs; very often the other species present also stunt. This results in a considerable loss of yield and economic profitability. Special measures must therefore be taken to overcome these problems. Two of the most common are (1) rearing a mixed male-female population of young tilapia and marketing them before they attain sexual maturity and (2) rearing all-male tilapia. These are received now mainly by manual sexing.

The factors that should be taken into account when deciding which method to employ are:

The Size Preference of the Market. If the market accepts small fish (below 250–300 g) which can be obtained before they reach sexual maturity, the first method should be preferred, since it saves effort, fingerlings, and space. If, however, the market demands only large fish which reach maturity before marketing it will be better to rear only males and thus avoid "wild" spawning and a stunted population. In some cases the market accepts both sizes, but prices are higher for larger fish. Careful economic calculations should then be made, weigh-ing the higher income from producing larger fish against the extra costs of sexing, and loss of females. The calculation should also con-sider that the yield is higher in an all-male population than in a mixed

TABLE 7.4 Monoculture (Common Carp or Tilapia) in Some Fish Farms in Israel

	Stocking Date	Harvesting Date	Rearing Period (days)	Density at Harvest (fish/ha)	Average Stocking wt. (g)	Average Harvesting wt. (g)	Yield (kg/ha)
Farm: Ginossar technological experiments station *Pond area: 0.02 ha*							
Tilapia (hybrid)	Aug. 20	Nov. 11	76	30,000	217	380	4890
						Daily gain (kg/ha): 64.3	
Farm: Nir David							
Tilapia (hybrid)	July 1	Sept. 1	62	18,400	250	450	3680
Common carp	July 1	Sept. 1	62	1,000	250	600	350
							4030
						Daily gain (kg/ha): 65.0	
Farm: Afikim *Pond area: 0.4 ha*							
Tilapia (hybrid)	Mar. 28	May 22	55	2,975	30	160	387
Common carp	Mar. 28	May 22	55	14,750	160	500	5015
							5402
						Daily gain (kg/ha): 98.2	

one since in most tilapia species the males grow more rapidly than the females (Fryer and Iles, 1972). In some species the difference in growth between males and females, and the higher yield obtained by culturing only males, justify the extra work involved in sexing or, if fry are bought, the higher price for all-male fry.

The Species of the Fish Reared. Some species, such as *Tilapia zillii* or *Sarotherodon mossambicus,* breed when they are small and cannot reach market size before maturation. Chimits (1955) states that reproduction of *S. mossambicus* can commence at an age of 2–3 months. Mironova (1969) found *S. mossambicus* spawning when only 2.5 g in weight. Since sexing can be done reliably only when the fish have reached a size of 50–70 g, they must be first nursed to this size. In many cases these species of tilapia breed at even a smaller size. Pruginin and Arad (1977) report that in Malawi, *S. mossambicus* bred and growth stopped due to stunting when the fish reached 30 g. As a consequence the yield during 100–150 days did not exceed 300 kg/ha. It may be advisable, therefore, to choose other species which spawn at an older age, such as *S. niloticus, S. aureus,* or others. Pruginin (FAO/UN, 1965) found that while *S. hornorum* in Uganda reached not more than 150 g in about 1 year, *S. niloticus* normally reached 250 g in 5–6 months, at which time they reached sexual maturation.

When rearing young-of-the-year unsexed populations, the time period before their sexual maturity is quite short, 3–6 months. The production plan should, therefore, be based on having two to three such cycles a year with complete draining of ponds between cycles. In Israel, because of the cold winter, there is only one rearing period from May to November.

In order to achieve market weight during a short period, stocking densities should be relatively low: 3000–5000 fish/ha. Recently hatched young fry should be stocked rather than stunted older fry which can spawn earlier. The importance of complete draining between rearing periods and elimination of the fish remaining in the drained pond cannot be overemphasized. If necessary, the fish remaining in pools must be poisoned, and the influx of fish from outside the pond should be prevented by screening the water inlet.

Due to the relatively low densities of fish, natural food is only partly utilized. Natural food can be better utilized through polyculture, with an appreciable increase in the overall fish production. The growth rats of tilapia is not affected by the presence of the other species if their density is not too high. Yashouv (1969) has demonstrated that 2500–3000 common carp or mullet per hectare did not inhibit growth of *S. aureus.*

Polyculture may have an additional advantage. Common carp (and grass carp), when large enough, can prey to some extent on tilapia fry (Spataru and Hepher, 1977). Thus introducing these carp into tilapia ponds can alleviate the problem of "wild" spawning that may develop at the end of the culture period. However, the problem of "wild" spawning can be much more efficiently solved by including a predator fish in the polyculture. The fishes used for this purpose were the nile perch (*Lates niloticus*) in Africa (Pruginin in FAO/UN, 1965; Meschkat, 1967); mud fish (*Ophiocephalus striatus*) in the Philippines (E. M. Cruz, personal communication) and in Thailand (Chimits, 1957); *Cichlasoma managuens* in Central and South America (Dunseth and Bayne, 1978); and sea basses (*Dicentrarchus labrax* or *D. punctatus*) in Israel (Chervinski, 1974, 1975). The last two are euryhaline marine fish. The use of predator fish in tilapia ponds did not receive wide application, however, and was practiced mainly on an experimental scale, partly because of difficulty in obtaining the predatory fish fry.

A typical example of rearing young-of-the-year *S. aureus* is given in Table 7.5 (from Halevy, 1979). He reared tilapia, which were

TABLE 7.5 Polyculture of Tilapia (*S. aureus*) with Common, Silver, and Grass Carp in a 1.4 ha Pond at Dor, Israel[a]

Fish Species	Stocked Density (fish/ha)	Stocked Avg. wt. (g)	Harvested Density (fish/ha)	Harvested Avg. wt. (g)	Gain Avg. Daily Gain (g)	Gain Daily yield (kg/ha)	Gain Total yield (kg/ha)	Annual Total Yield (kg/ha)
First Period:	February 6–June 30							
Tilapia	5000	6	2140	200	1.3	2.8	398	
Common carp	3000	5	2960	642	4,4	13.0	1885	
Silver carp	930	378	890	2000	11.3	10.2	1428	
Grass carp								
Second Period:	July 3–November 22							
Tilapia	5000	0.5	4560	224	1.6	7.3	1018	1416
Common carp	7170	125	4450	533	2.9	12.9	1776	3661
Silver carp	1070	750	1070	2220	10.3	11.0	1572	3000
Grass carp	1430	10	1290	150	1.0	1.3	180	180
			Annual grand total:					8257

[a] From Halevy (1979).

hatched in May, during the second growth period starting in July. Average fry weight at stocking was 0.5 g. They were reared until November, when they reached 224 g. During the first growth period (February to May) Halevy (1979) reared unsexed tilapia fingerlings hatched in the previous year. This was possible since this season in Israel is warm enough to allow tilapia growth but not spawning (spawning occurs only when water temperature reaches above 20°C). This last method of rearing mixed populations of tilapia hatched during the previous year is not practical for commercial farms. Since the stocking and harvesting of ponds on those farms are phased so as to meet a marketing schedule (see Section 7.4.2), it is hard to limit the growing period to the cool spring. When this period is extended to the summer, spawning will start. Also, since overwintering of tilapia fingerlings is costly (Avault et al., 1968), it is more economical to overwinter sexed male tilapia which will grow in spring much better than the mixed male-female population. Thus, young-of-the-year tilapia are cultured on commercial farms in Israel during the second half of summer. During the spring an all-male tilapia population is usually cultured.

Culture of an all-male tilapia population remove the restrictions imposed on the final age of the harvested fish. The fish can be cultured to an older age and thus larger final weight than young-of-the-year. A weight of 400–600 g can be easily obtained during 1 year. Since rapid growth is not as important as in the culture of young-of-the-year, the fish can be stocked at higher densities. Growth rates may be reduced, but yields per unit area usually increase. Fish farmers then have the choice of culturing tilapia in a polyculture system at relatively low densities of 3000–5000 fish/ha (see Section 7.3.1), or at high densities of 10,000–20,000 fish/ha, and sometimes even more, in monoculture (see Section 7.3.2).

When sexing is employed, it is important to sex the fish carefully. The fewer errors made in sexing, the less the likelihood of "wild" spawning. The sexing of males and females of most tilapia species is relatively simple. In most cases the tilapia can be distinguished by the genital papilla, which has one orifice in the male as compared to two orifices in the female (Figure 7.5). The female often also has a smaller genital papilla. There is some conflict about the time of sexing. The earlier the sexing is done, the better, since the females are discarded and early sexing saves space which can be used for rearing of males. However, there is a certain minimum size at which sexing can be accomplished with an acceptable degree of confidence. Even under the best conditions and when done by trained people there is a certain

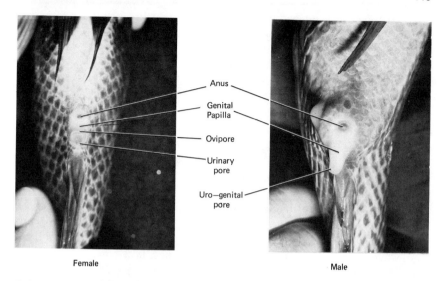

Anus

Genital
Papilla

Ovipore

Urinary
pore

Uro–genital
pore

Female   Male

FIGURE 7.5. Secondary sex characteristics of *Sarotherodon niloticus*. (Courtesy of S. Rothbard, Dor, Israel.)

error in sexing, usually up to 3–4%, as a result of which some "wild" spawning occurs in rearing ponds.

The tedious work of sexing and the precautions mentioned above are not necessary when monosex fry are used. There are obtained either by crossing certain species of *Sarotherodon* or by sex reversal treatments (see Section 6.2). However, while these methods appear promising, no commercial application on a large scale has yet been made of them.

An important consideration in deciding whether to employ manual sexing or hybridization is the hazard to natural tilapia populations of introduction an exotic species which may compete or cross with the natural population, thus upsetting the natural balance in natural waters. For instance, *Sarotherodon hornorum* by itself has little value because of slow growth and a dark color which does not appeal to consumers in many regions. It is, however, a valuable fish for hybridization since most of its crosses produce all-male populations. The decison on whether to import this fish should depend in part on the relative importance of wild tilapia fisheries as compared to pond culture. Where pond culture is more important, the use of hybridization and introduction of exotic species for this purpose may be justified. In some African countries natural tilapia fisheries are important and so such introductions should be examined critically. In Israel, the possible affected species in natural waters is *S. aureus,* the production of which is only 200 ton/yr as compared to 2500 ton/yr of this species cultured in ponds.

7.3.4 Management of Nondrained Ponds

Problems associated with ponds that are not drained have been discussed earlier (see Section 1.2). Still, in many parts of the world, especially South Asia, such ponds are common. To obtain the maximum possible yields from such ponds special attention should be given to a number of management practices that can alleviate the problems involved.

Efforts should be made to drain the ponds, if possible. When the only obstacle to draining is the lack of a gravity outlet, pumping the water out may be a proper solution. A number of portable pumps have been developed for this purpose. In many cases, however, the difficulty is not the lack of technical facilities for draining, but rather the lack of water to refill the pond after draining. If possible, a suitable water source, such as a borehole well (tube well) should be found in the immediate vicinity of the pond. Another solution is to construct two or more ponds and pump the water from one pond for temporary storage in the other. A connecting pipe with a monk in either pond may be a great help in transferring water. If draining (by pumping) is not possible every season or every year, then draining every few years, some time before the rainy season starts, is advisable. This will aerate the bottom soil and help alleviate problems occurring in the pond.

Before each rearing period wild fish and other pests should be eliminated. Rotenone is a common piscicide used for this purpose. It should be applied so as to reach a concentration of 1 ppm. If it is not available, other piscicides should be used. Ghingran (1975) mentions a number of piscicides such as the seed powder of *Barringtonia acutangulata* which kills a wide variety of fish at a concentration of 20 ppm, or Mahua oilcake (*Bassia latifolia*) which kills fish at a concentration of 60–100 mg/1. These plants are common to India and some Southeast Asian countries. The poisonous effect of these powders lasts for only 48 hours. Being organic matter, these piscicides also serve as good manure for the fishponds.

If water other than rainwater enters the ponds, proper screening against wild fish should be employed (see Section 4.4.2). Predatory fish, or other predators such as frogs and snakes, may be of major concern in nondrained ponds since they may reduce the number of stocked fish. If the entry of predatory fish cannot be completely avoided by screens, the maximum weight limit of prey should be ascertained. Fingerlings should be stocked only when they are larger than this limit. Up to this weight the fingerlings should be nursed in specially controlled ponds.

In nondrained ponds it is best to grow, in polyculture, fish that cannot breed spontaneously. In addition to the biological balance of the species combination in the polyculture, special attention should be given to inclusion of fish species that improve environmental conditions. Fish feeding on macrophytes (such as grass carp), detritus (such as grey mullet—*Mugil cephalus* where available; or mrigal—*Cirrhinus mrigala*), molluscs (black carp—*Mylopharyngodon piceus*), and phytoplankton (silver carp or catla—*Catla catla*) will reduce the amount of organic matter and excess algae, improve conditions, and increase yield of such ponds. When there is an intrusion of small wild fish of low or no commercial value, or if tilapia are included in the species combination, it is recommended that a predatory fish be included. The predator will control the intruding fish and wild spawning. Such a predator should be relatively small so that it will not prey on stocked fish. It is also prefarable that this fish does not breed spontaneously in the pond (an example is the sea bass, *Dicentrarchus* sp.).

In order to better regulate fish density, harvesting methods that efficiently remove most of the fish should be employed. A small number of fish which remains will probably be caught during the following harvest. However, if their number is high the growth rate of the fish stocked in the following growth period may be affected.

7.4 ANNUAL PLANNING

Profitability of the fish farm depends not only on proper choice of management methods, but also on the most efficient use of resources. This requires advance planning of fry production or purchase and planning of the production schedule in the rearing ponds.

7.4.1 Planning of Fry Supply

The number of fry required on a farm is determined by the total number of fish marketed during the year and the losses incurred at each stage of rearing and nursing. Supplies should be calculated separately for each species reared.

Table 7.6 presents an estimate of the number of fry required at the beginning of each stage in order to get 1 kg of fish to market. The loss at each stage is the expected average and may vary under different conditions. The average weight of market fish may also vary, in which case the number of fish required has to be corrected accordingly. If fry

TABLE 7.6 Number of Fry Required for Stocking, at Each Stage of Culture, to Produce 1 kg of Market-Size Fish (Israeli Conditions)

Species	Market Size (g)	No. per 1 kg	Rearing			Nursing II			Nursing I		
			Weight Range	Expected Loss (%)	Fry No.	Weight Range	Expected Loss (%)	Fry No.	Weight Range	Expected Loss (%)	Fry No.
Common carp	600	1.7	100–600	6	1.8	0.2–100	50	3.6	—	—	—
Tilapia, mixed population	200	5.0	3–200	30	7.1	—	—	—	—	—	—
Tilapia, all males, sexed from 50% males	500	2.0	70–500	10	2.2	70[a]	55[a]	4.9	1–70	30	7
Tilapia, all males, sexed from 90% males	500	2.0	70–500	10	2.2	70	30	3.2	1–70	30	4.5
Mullet, 1st yr.	400	2.5	1–400	30	3.6	—	—	—	—	—	—
Mullet, 2nd yr.	600	1.7	100–600	20	2.1	0.3–100	50	4.2	—	—	—
Silver carp	1500	0.7	200–1500	7	0.8	25–200	10	0.9	2–25	30	1.2

[a] Mixed population before sexing and wintering. The loss at this stage is a combination of 50% discarded females and 10% loss of males during wintering.

are purchased from outside the farm, their number will depend on size and stage of nursing or rearing, as specified in Table 7.6.

The age at which fry are stocked is often important. In warm climates fish attain sexual maturity earlier than in colder climates. In Israel carp reach sexual maturity when 1 year old. It is obvious that it is better not to allow them to reach this age in the rearing ponds or, if they do, it should be late in the rearing period before harvesting. Fry spawned then are too small to compete for food. However, this does pose some problems. Since the natural spawning period of carp in Israel is April to May, nursed fingerlings are not ready for stocking in rearing ponds before June. If during the early part of the growing season (March to June) fingerlings spawned the previous April or May are stocked, they will reach 1 year of age during the rearing period and usually spawn. Fingerlings from late season spawning must, therefore, be used (see Section 6.1.3).

"Wild" spawning becomes more acute with the Chinese "big belly" variety of the common carp, which breeds much earlier than the European common carp (see Section 5.1). It should be noted here, however, that in the tropics carp spawn year round, and there is no problem in obtaining fry or fingerlings during any season.

No less problematic is the age of tilapia cultured in the ponds since they breed at an even earlier age than carp. This becomes critically important when mixed populations of males and females are reared (see Section 7.3.3). Since tilapia are prolific, the population of fry in a spawning pond may consist in part of older, stunted fingerlings. These will spawn earlier and at a smaller size than newly hatched tilapia fry. Therefore it is important to drain the spawning pond of tilapia relatively often, arrange for new spawning, and use young fry or fingerlings.

7.4.2 Planning Production

The yearly plan of production should give a tentative but rather detailed scheme of pond allocation for the various production activities. The production plan should designate expected average fish weights at stocking and harvesting, where the fry or the stocked fish come from, expected yield and how it complies with production targets, requirements for production (such as supplementary feed, fertilizers, and manures), and, finally, a tentative timetable (chronogram) of activities for each pond. A good production plan should also provide an estimate of expected profit and try to optimize it.

Since the production plan is determined by a set of conditions such as the amount of fish that can be produced or marketed, the available fry, pond productivity, and other factors, and these do not change rapidly, the framework of the yearly production plan usually remains fairly constant, changing only gradually from year to year with changes in consumers' demands (species and amounts), the availability of production inputs, and the introduction of new production methods and programs. The yearly plan of production should, however, be flexible enough to adapt to unforeseen sudden changes in conditions such as the availability of water, fry, feed, and other items.

The first step in planning yearly production is to set production targets. These are determined by any one or combination of the following factors: (1) the availability of fry, (2) the production or handling capacities of the farm, and (3) the marketing potential of each species. Limitations here, if they exist, may be imposed as marketing quotas, as is the case in Israel where the Fish Breeders' Association regulates the market.

Once production targets are determined, it is necessary to calculate the number of fry of each species required to meet the targets. The fry may be purchased, if available, from outside sources. They can be in stock from the previous year; or they can be produced on the farm during the production season and, as in most cases, nursed in the farm. Ponds should be allocated both for spawning and nursing.

The need to rear some fish in two stages—the first stage (in Israel this is usually in the first year) to an intermediate size and then restocking to be reared to market size (in Israel this is usually in the second year)—calls for the employment of various culture programs. Some programs provide fish for market, while others provide fish for restocking. It is obvious that even if fry, nursed fingerlings, or first stage fish are available at the beginning of the year in sufficient numbers for stocking, the need to spawn, nurse, and rear is not diminished, since similar stock must be prepared for the following year.

Planning production is rather complicated and has to take into account a number of factors simultaneously. In order to elucidate the procedure, the following example is given. It involves production for a farm of 50 ha. This example is a modification of an actual production plan for a farm of similar size in Israel. The farm in our example is divided into the following ponds:

Ponds A_1–A_3	0.1 ha each	=	0.3 ha
Ponds A_4–A_5	0.5 ha each	=	1.0 ha
Ponds A_6–A_{10}	1.0 ha each	=	5.0 ha
Pond A_{11}	2.0 ha		2.0 ha
			8.3a ha

Total area of auxiliary ponds:
(about 17% of total farm area)

Pond P_1	2.0 ha	=	2.0 ha
Pond P_2	2.2 ha	=	2.2 h
Ponds P_3-P_{17}	2.5 ha each	=	37.5 ha
Total area of production ponds:			41.7 ha

The production targets of the farm were set as follows:

Common carp	110 tons/yr	(avg. wt. 600 g)
Tilapia	60 tons/yr	(avg. wt. 500 g)
Silver carp	20 tons/yr	(avg. wt. 1500 g)
Mullet	10 tons/yr	(avg. wt. 500 g)
Total production:	200 tons/yr	

This production target was determined primarily by the production potential of the farm as ascertained from previous years. In the present case this was quite high. The data presented show that the expected average yield is about 4.0 ton/ha/yr, or if the auxiliary ponds are excluded, 4.8 ton/ha/yr. The ratios among species were determined in part by market demand and also, as in the case of mullet, by fry availability.

The requirements of fry and nursed fingerlings and the area needed for producing them are then calculated (Table 7.7). The following assumptions were made in preparing the table:

Common carp	All carp are produced in "late season" spawning. In case of spawning failure, spring spawning will be done. The fry will be nursed during autumn to 10 g and during winter and early spring to 100 g.
Tilapia	The 500 g tilapia will be produced by rearing only males in their second year. These will be obtained by sexing a normal-to-female population in autumn when the fish reach 70 g and by wintering the males.
Silver carp	Fry are purchased at an average weight of about 1 g and stocked into rearing ponds. The expected loss before obtaining 400 g average fish at the end of the year is about 40%.
Mullet	Fry captured in natural bodies of water are held during winter in holding ponds are then nursed to a weight of about 1–2 g. These are stocked into polyculture ponds and reach 200 g by the end of the year.

TABLE 7.7 Fry Required to Attain Production Targets for the 50 ha Farm Given as an Example in the Text and the Area Required for Spawning and Nursing[a]

Requirements for Rearing Ponds

Species	Production Targets (ton/yr)	Average Stocking Wt. (g)	Number Required
Common carp	110	100	200,000
Tilapia	60	70	(sexed) 132,000
Silver carp	20	400–500	16,000
Mullets	10	200–300	21,000

Production in Spawning Ponds

Species	Brood ♀	Stock ♂	Area Required (ha)	Time	No. of Expected Fry
Common carp	30	45	0.7	July–Aug.	300,000
Tilapia	210	60	0.3	May–June	420,000

Requirements for Nursery Ponds

Species	Size Range (g)	Density (fish/ha)	Area Required (ha)	Time
Common carp	2–10	100,000	3.0	Sept.–Oct.
Common carp	10–100	10,000	20.0	Nov.–Mar. 15
Tilapia	1–70	100,000	3.0	July–Oct. 15
Tilapia (sexed, during winter)	70	500,000	0.3	Oct.–Mar. 15
Mullet	0.3–1	10,000	0.4	Nov.–April

[a] See also Table 7.5.

Allocation of auxiliary ponds for spawning and nursing is also presented in Table 7.7.

The main culture program employed on the farm during the spring was polyculture of four species as shown in Table 7.8. This combination was found to be suitable to the particular conditions on this farm and somewhat different from the polyculture combinations given in Section 7.3.1.

Since the number of mullet fry is limited and not sufficient to stock all the rearing ponds, this combination is employed only in half of the available ponds. On the other hand, because it is necessary to nurse mullet for further rearing the following year, a somewhat different combination (II) is used (see Table 7.9) in which mullet fry of 1–2 g

TABLE 7.8 Polyculture Combination I (Spring—120 Days)

Species	Avg. Wt. at Stocking (g)	Stocking Density (fish/ha)	Expected Harvest Wt. (g)	Expected Harvest (ton/ha)
Common carp	100	3000	600	1.8
Tilapia	70	4000	250	1.0
Silver carp	500	300	1500	0.45
Mullet	200	1000	500	0.5

TABLE 7.9 Polyculture combination II (Spring—120 Days)

Species	Aver. Wt. Stocking (g)	Stocking Density (fish/ha)	Expected Harvest Wt. (g)	Expected Harvest (ton/ha)
Common carp	100	3000	600	1.8
Tilapia	70	4000	250	1.0
Silver carp	500	300	1500	0.45
Mullet	1–2	5000	100	0.4

each are stocked at a density of 5000 fry/ha instead of year old mullet. These reach about 100 g by July. At this density about 25% of the rearing pond area is sufficient to produce the necessary mullet for stocking the next season. The rest of the ponds are stocked by polyculture combination III (see Table 7.10) which does not include mullets at all. Except for the size of mullet or their presence, combinations II and III are equal to combination I.

The main culture program employed in the second rearing period starting in July/August is shown in Table 7.11. Here again, however, the number of mullet fry is not sufficient to stock more than about 25% of the area so modification of this combination (V—see Table 7.12) does not include mullet fry.

TABLE 7.10 Polyculture Combination III (Spring—120 Days)

Species	Aver. Wt. at Stocking (g)	Stocking Density (fish/ha)	Expected Harvest Wt. (g)	Expected Harvest (ton/ha)
Common carp	100	3000	600	1.8
Tilapia	70	4000	250	1.0
Silver carp	500	300	1500	0.45

TABLE 7.11 Polyculture Combination IV (120 Days)

Species	Aver. Wt. at Stocking (g)	Stocking Density (fish/ha)	Expected Harvest Wt. (g)	Expected Harvest (ton/ha)
Common carp	100	1500	600	0.9
Tilapia	250	3000	500	1.5
Silver carp	1–2	1200	500	—
Mullet	100	2000	200	—

TABLE 7.12 Polyculture Combination V (Summer—120 Days)

Species	Aver. Wt. at Stocking (g)	Stocking Density (fish/ha)	Expected Harvest Wt. (g)	Expected Harvest (ton/ha)
Common carp	100	1500	600	0.9
Tilapia	250	3000	500	1.5
Silver carp	1–2	1200	500	—

The rearing of silver carp is also limited. In order to supply the market demand it is sufficient to rear the fry during the second rearing period for "second stage" fish in the following year in only 50% of the rearing ponds. The rest of the area is therefore stocked with carp and tilapia only (polyculture combination VI—see Table 7.13).

Tables 7.14 and 7.15 present in schematic form the allocation of auxiliary and rearing ponds during the year and their expected production. This plan only gives averages for fish weight and dates of stocking and harvesting for groups of ponds. The time required for draining, harvesting, refilling, and restocking the ponds and the limited manpower for carrying out these tasks make it necessary to spread the time of stocking and harvesting among the various ponds. Deviations from the average times given in the plan are naturally also related to harvested fish weights. A more detailed production plan should take account of these factors.

TABLE 7.13 Polyculture Combination VI

Species	Aver. Wt. at Stocking (g)	Stocking Density (fish/ha)	Expected Harvest Wt. (g)	Expected Harvest (ton/ha)
Common carp	100	1500	600	0.9
Tilapia	250	3000	500	1.5

TABLE 7.14 Allocation of Auxiliary Ponds for the Different Culture Activities During the Year on the 50 ha Fish Farm Given as an Example in the Text

Pond	Area (ha)	J	F	M	A	M	J	J	A	S	O	N	D
A_1	0.1	← Segregation and holding pond for carp brood stock →											
A_2	0.1	← Segregation and holding pond for carp brood stock →											
A_3	0.1	← Holding pond for tilapia brood stock →											
A_4	0.5	← Nursing mullets →				Tilapia spawning →		Carp spawning →				Nursing mullets	
A_5	0.5	← Tilapia wintering →						Carp spawning →				Tilapia wintering	
A_6	1.0	← Storing tilapia for marketing →						← Tilapia nursing →				Storing tilapia	
A_7	1.0	← Storing carp for marketing →						← Tilapia nursing →				Storing carp	
A_8	1.0	← Storing carp for marketing →						← Tilapia nursing →				Storing carp	
A_9	1.0	← Carp nursing II and holding →								Carp nursing I →		Carp nursing II	
A_{10}	1.0	← Storing silver carp for marketing →						← Storing silver carp for marketing →					
A_{11}	2.0	← Carp nursing II and holding →								Carp nursing I →		Carp nursing II	

MONTH

TABLE 7.15 Allocation of Rearing Ponds for the Different Culture Combinations During the Year on the 50 ha Fish Farm Given as an Example in the Text

MONTH

Pond	Area (ha)	J	F	M	A	M	J	J	A	S	O	N	D
P_1	2.0	—Carp nursing II→ ←Polyculture combination I→ ←Polyculture combination IV→											
P_2	2.2	←Polyculture combination I→ ——Storing carp for marketing——											
P_4–P_8	2.5	—Carp nursing II→ ←Polyculture combination I→ ←Polyculture combination V→											
P_9–P_{13}	2.5	←Polyculture combination III→ ←Polyculture combination IV→											
P_{14}–P_{18}	2.5	←Polyculture combination III→ ←Polyculture combination VI→ ←Carp nursing II→											

Yield

Production of market fish

Common carp	←————35.5 tons————→
Silver carp	←————75 tons————→
Tilapia	←————18.75 tons————→
	←————60 tons————→
Mullet	←————8.35 tons————→

Deviations from the plan can be caused by variations in climate or any other factor that affects fish growth rate. The plan as presented is based on production during eight summer growing months and marketing throughout the year. This requires the allocation of rather large areas for storage of marketable fish. In warmer climates production can be much better regulated and spread more evenly over the year, thus reducing the area of storage ponds.

8 ⎯⎯⎯⎯⎯⎯⎯⎯⎯⎯⎯⎯ Techniques
and Equipment

8.1 ROUTINE WORK

One of the most important routine operations, and one which should be done every morning, is the checking of pond condition for any irregularities. These might be technical, such as water breaking through an embankment, blocking of water supply, and other problems, or they may relate to the fish behavior. When fish are surfacing and gulping air, it is typically a sign of anoxia. When they concentrate near the shores, appear very weak, and exhibit poor equilibrium, it may be poisoning by *Prymnesium parvum* (see Section 11.2) or by other toxicants coming from outside the pond. Irregularities also can be observed in the color of the water. They can be caused by a change in the composition of the plankton due to a serious imbalance in the biological equilibrium of the pond. The appearance of a bloom of *P. parvum* turns the color of the water brown, while a mass mortality of the phytoplankton due to climatic conditions or the lack of essential nutrients usually appears as a whitish color. It is also important to observe whether seagulls are concentrated in certain ponds, since they point to a weakness or disease of the fish in those ponds. Seagulls recognize signs of stress in fish long before the fish farmer.

It is important to check pond condition as early as possible in the morning. The early morning hours are critical from the point of view of monitoring the pond's oxygen regime (see Section 11.1). During these hours, if anoxic conditions exist, and the fish show signs of stress, it may be too late for correction since the damage to the potential growth of the fish has already occurred. Many fish farmers are therefore now using an oxygen meter. Oxygen is measured as a matter of routine in every pond. In some farms the concentration of oxygen is also measured in the late afternoon in ponds suspected of anoxic conditions. If the oxygen concentration is low at that time, there is a danger that it will not be sufficient for fish needs during the night. Therefore special measures must be taken to correct the problem, such as increasing the water flow or providing aeration.

During the examination it is important to inspect feeding sites to verify that all the feed given the previous day has been consumed. If unconsumed feed is found, feeding should be stopped and the reason for this condition determined. The fish may stop eating for a number of reasons, such as low temperature, disease, stress due to anoxia, and overfeeding. In spring, the cause could be an abundance of natural food to the point that supplementary feed is not required or can be significantly reduced.

Ponds must be protected from ingress of "wild" fish, especially where these are numerous in the source from which the pond receives its water. Intruding fish may propagate, compete for food, and lead to stunting of the whole population. They can also contribute to parasite and disease problems. *Tilapia zillii* was such a pest for many years in Israel. The introduction of *Sarotherodon aureus* has solved this problem almost entirely.

In some places, carnivorous fish may enter ponds and devour stocked fry (Woynarovich, 1975). Where intruding fish are a problem, they must be screened off. These screens may be of permanent construction, such as a box constructed in the inlet canal made of earth, bricks, or concrete and filled with gravel (see Section 4.4.2). Or the screens can be devices of a more temporary nature such as screening boxes or sleeves (see Figure 8.1).

A screening box is placed under the inflow water. The box must be strong enough to withstand battering by the water. A box with a bottom of sheet metal and walls of strong screen will be quite efficient in preventing the ingress of fish. However, such a box is usually quite costly. A much simpler method is a screening sleeve made of synthetic

FIGURE 8.1. A screening sleeve on an inlet pipe. Note that the inlet pipe is near the outlet monk.

fibers (saran). The length of the sleeve is determined by the amount of water and its cleanliness. In severe cases the sleeve should be about 2 m long (Pruginin, 1956). The sleeve is closed tightly at its end.

Both the box and the sleeve must be cleaned from time to time. With the sleeve this is done by opening its far end and emptying the contents into a box or pail.

If wild spawning of carp occurs, the number of fry produced can be reduced by spraying a solution of Rotenone or any other piscicide along the banks of the pond during the first few days after first hatching. During this time the fry (or eggs of later spawns which have not yet hatched) are concentrated in the shallow areas along the banks. Spraying these areas can reduce the number of fry and their potential damage. Larger fish are not affected as they avoid these areas, but the fry cannot leave for the first 2–3 days and succumb. The dosage is 1 l of concentrated Rotenone solution to 1000 m of bank length. The Rotenone is diluted first with water and then applied with a low-pressure sprayer. Since not all fish spawn simultaneously, the spraying should be repeated a number of times at intervals of a few days. Some farmers apply this treatment prophylactically in ponds where the danger of "wild" spawning exists. It is done weekly from the time that the first spawning is noticed and continues for 3–4 weeks.

Chemical treatment is not efficient against wild spawning tilapia. The tilapia species cultured in ponds are mouth brooders, and the fry are protected in their mother's mouth for the first days of their life.

It is important to eradicate all fish remaining in pools and puddles immediately after draining and harvesting. This is especially important with fish that reproduce easily, such as tilapia, and in ponds that cannot be completely drained. Eradication is usually done with Rotenone, which is sprayed so as to create a concentration of 1 ppm in the pools. All pools must be sprayed. After a few minutes the fish float to the surface. If it is desired to revive them, they should be immediately collected and put into a tank containing a solution of 1 ppm potassium permanganate which oxidizes the Rotenone and counteracts it. The fish will then revive quickly. In some countries, such as the United States, Rotenone has not been cleared for use with food fish, and it is illegal to market fish that have been killed with this chemical.

When Rotenone is not available and the pools are not too large, ammonium sulphate fertilizer can be used. A generous dose should be cast into the pools.

The eradication of fish remaining in the pools permits refilling of the pond after only a short period. Before refilling, however, the sluice boards of the monk should be checked, small repairs in the embankment made, if required, and the filtering devices checked and repaired or replaced.

8.2 FEEDING

Diet composition and feeding rates are discussed in Chapter 10. Here we shall consider actual feeding methods. Feeding is done soon after the morning check. It can be carried out manually by casting the feed from a boat, or a tank from which feed is blown into the pond can be used. Since both methods demand considerable time and effort, feeding is usually done only once a day. Feeding also can be carried out continuously by mechanical feeders.

8.2.1 Feeding Manually or from a Blower Tank

Feeding is usually done in a fixed place in the pond, if possible in the harvesting sump, so that the fish will stir up the silt and help clean the sump. For the daily check for leftover feed, the feeding place is marked with a pole. If the feeding place is in deep water, the check can be done from a boat with a "shovel" made of wire screen fitted to a long pole. If the feed has not been eaten, feeding must be stopped until previous feed has been completely consumed. If, however, signs of decaying feed appear, the feeding place should be changed since the fish will avoid this place.

The manual work involved in feeding can be eased to a large extent by the use of a blower tank. A typical blower tank contains 2–4 tons of feed. It is equipped with an impeller which blows the feed through a slanted pipe a distance of about 10 m into the pond. The tank can be mounted on a truck, and the blower propelled by the motor of the truck, or it can be a single-axle tank pulled by a tractor and propelled by its power takeoff (see Figures 8.2 and 8.3). The tank has a double bottom slanted toward the center. In the lowest part of the tank, in the center, there is an auger which transfers the feed to the blower at the front or at the rear of the tank. Since each turn of the auger transfers a constant amount of feed to the blower (depending on the kind of feed), the amount of feed applied to the pond can be measured by counting the number of turns of the auger. For this purpose a meter which counts the rotations of the auger is installed, either in the driver's cabin of the car or on the tank where it can easily be read by the tractor driver. Feeding from a blower tank is also done in a fixed place in each pond, usually in the harvesting sump. Very often the fish

FIGURE 8.2. A single-axle feeding blower tank pulled by a tractor and operated by its power takeoff. (Courtesy of S. Sarig, Nir-David, Israel.)

recognize this place and concentrate there when they hear the noise of the vehicle approaching.

8.2.2 Mechanical Feeders

These can be divided into two groups: (1) demand feeders and (2) automatic feeders. Demand feeders are activated by the fish by means of a rod which hangs from the feeders into the water. Movement of the rod releases a small amount of pellets. Automatic feeders are activated by electric timers set at predetermined intervals to release feed for predetermined time periods.

FIGURE 8.3. A feeding blower tank mounted on a truck. (Courtesy of S. Rothbard, Gan-Shmuel, Israel.)

Demand Feeders

The main advantages of demand feeders over automatic feeders are (1) their low price and (2) the fact that on small farms, or in small ponds, they can be constructed from materials found on every farm— materials such as oil drums, pipelines no longer usable for irrigation, and other items. Unlike automatic feeders, there is no need for an electrical supply. This is important as some ponds are far away from a source of electricity. Another important advantage is that demand feeders provide feed only when activated. If for any reason (such as disease or anoxia) the fish do not feed, feed is not spent. This has special importance during periods of anoxia since unconsumed feed in the water undergoes bacterial decomposition and aggravates the anoxic condition in the pond. The main disadvantage of demand feeders is the difficulty in regulating the amount of feed dispensed, either because of blockages in the openings or because the openings are too large.

On some farms feeding by demand feeders is ad libitum and the fish can eat as much as they want. According to some fish farmers who use this technique, feed conversion ratios are not higher than usual. Other fish farmers claim that ad libitum feeding does increase the feed conversion ratio and, therefore, the daily dose provided to the fish is limited, though feeding time is spread over a longer period of the day.

The demand feeder can be very simple: any container with a conical bottom and having an opening of about 5 cm diameter in the center. Through the opening hangs an iron rod 12–14 mm thick. On the rod, below the opening, a ball or an inverted cone is mounted so that it almost blocks the opening. The gap between the ball and the opening can be regulated. When the rod is moved to any side by the fish, this gap increases and feed falls down into the water. The rod hangs to 20–30 cm below water level. The fish quickly learn how to move the rod and activate the feeder. Many other designs of demand feeders exist. The one described has proven itself in field conditions. Carp larger than 200–250 g are the best users of demand feeders. Small carp in nursery ponds or tilapia cannot activate the feeder themselves and, therefore, a number of larger carp are stocked in these ponds in order to activate the feeder and release feed for the other fish.

The opening in the feeder should be at least 70 cm above the surface of the water, since water splashed during intense feeding may otherwise reach the feeder, wet the feed, and block the opening. This is why an inverted cone, as a check for the dispensing of feed, is better than the ball. The cone protects the opening against splashing. The container itself can be of any kind; for example, an oil drum (Figure 8.4) of 200 l capacity, the bottom of which has been replaced by a cone,

FIGURE 8.4.   A demand feeder made of a used oil drum. (Courtesy of S. Sarig, Nir David, Israel.)

can be used for a pond of about 1 ha (or a number of drums for larger ponds). It is advisable to construct the drum on a swinging arm so that although it is at a distance of about 5 m from the bank, it can be swung in to the bank for refilling. Note that because of the large weight and resulting momentum at the end of the arm, it is necessary to moor the arm well.

Another possibility is the use of an aluminum pipe with a diameter of 6–8 in. This is installed diagonally so that the lower end is over the pond. At this end a cone bottom with an opening is fitted. The upper end is on the embankment. A 6 m long pipe contains about 150 kg of feed.

These relatively small containers demand daily refilling, and this is usually done from a blower tank. A flexible pipe is added to the outlet of the tank, and this pipe is placed in the containers during filling. To avoid refilling containers every day, a large hopper of 3–12 tons, built on the embankment or directly over the water, may be used (Figure 8.5). This hopper is filled once every several days directly from the feed distribution truck of the feed mill. A number of pipes (usually two or three) about 13 cm in diameter extend from the bottom of the hopper, and at the end of each is a funnel as described above. These larger feeders are more suitable for large ponds. They can serve for ad libitum feeding or, if a valve is installed between the hopper and the pipe arms, they can be filled with a certain amount of feed so that the daily rate can be controlled.

FIGURE 8.5. A large demand feeder over a pond.

Automatic Feeders

Automatic feeders distribute the feed from a hopper which is built on the embankment, usually near a corner of the pond, with the aid of a springlike auger inside a 5 cm (2 in.) aluminium pipe, the bottom of which (in the part passing over the water) is perforated at intervals of about 1 m by holes 8–10 mm in diameter. These holes permit passage of the pellets which fall through them into the water. The pipe usually passes across the corner of the pond so that the end farthest away from the hopper reaches the embankment at the other side. Here it is connected to an electric motor which turns the auger at fixed intervals and for fixed time periods according to a timer which controls it. When the feeding pipe is installed across the corner of the pond, it is best, if possible, to avoid having supports in the water, so that they do not interfere with fish sampling (see Section 8.3). See Figures 8.6 and 8.7.

The hopper of the automatic feeder is usually quite large, 4–10 tons. One hopper can serve two adjacent ponds. This is made possible by passing the feeding pipe below the surface of the embankment. Again, many other designs of automatic feeders are in use. Some blow feed into the pond at intervals; others have an automatic hatch. The one described was found to be most satisfactory in field conditions.

The advantage of the automatic feeder is the control it gives over the amount of feed added to the pond. Its disadvantages, however, are:

FIGURE 8.6. An automatic feeder. Note the electric timer near the hopper and the concentration of fish near the feeder pipeline.

1 High costs of construction.

2 The need for an electrical supply network in the pond area.

3 A potential waste of feed since the feeder operates whether the fish are feeding or not. This is important, especially in transition seasons at the beginning of the feeding period in the spring and at its end in the fall.

Automatic feeders, or a combination of automatic and demand feeders, have recently been developed which solve, at least in part, the last problem. Associated with these feeders are sensors which stop the action of the timers and feeders if fish do not come to feed within a

FIGURE 8.7. Fish concentrating and feeding below the automatic feeder's pipeline.

certain period of time (approximately 1 minute). Another arrangement is the installation of a small demand feeder which is filled from a larger hopper at periods controlled by a timer. If the fish do not come to feed, the small demand feeder is filled to its capacity, and this activates a microswitch which stops any further supply of feed from the hopper until the demand feeder is empty.

8.3 SAMPLING

One of the most important means for proper management of the ponds is the regular sampling of fish. Sampling allows calculation of feeding levels and provides a check on the condition and rate of growth of the fish. On commercial farms using intensive management methods, the feed is one of the largest operational costs. It is very important to adjust the amount of feed used according to fish size so as to obtain maximum possible growth or avoid overfeeding. This can be done only through frequent sampling of the fish and determining their average weight. Without routine sampling of every pond at intervals of 1 week to 10 days, there is no possibility of managing a fish farm in a profitable way. Sampling is usually done by catching the fish with a seine at the feeding place. If feeding is done at the harvesting sump, then since this place is too deep for seining, some feed should be added at a shallow corner of the pond a day or two before sampling is to be done so as to attract fish to that place. On the day of sampling, and 15–30 minutes after feeding, the designated area is encircled with a seine and the fish are netted (Figure 8.8). This should be done quickly so as not to give the fish time to escape. The larger the sample the more reliable it will be in representing the total fish population. In a regular rearing

FIGURE 8.8. Sampling fish by seining part of the pond.

pond the sample should be at least several hundred fish. The fish are counted and weighed, and their average weight and the growth rate from the previous sample are calculated. Each fish species is counted separately, and the results are recorded on the pond's card (see Section 8.6). These data serve as a check on the growth and food utilization in the preceding period as well as for calculating the feeding rate in the succeeding period until the next sampling. During sampling the fish are examined with respect to condition and health (diseases, parasites, etc.).

Sampling is usually done with a seine of 25 mm mesh, 20–30 m in length, and with a height of 3–4 m, so that fish of all sizes (over 25 g) can be caught. Sampling is done by two or three people, who pull the net from its ends. At each end there is a wooden brail pole between the lead and float lines. These help spread the net and serve both for pulling the net and pushing the lead line toward the bottom of the pond. The net is carried on a special stretcher. This stretcher can be made to float by making it out of a double board. One or two inflated car inner tubes are inserted between the boards. Dropping the net from a floating stretcher is much quicker than pulling it around the sampling site. This is important to avoid escape of fish during encircling of the site.

When the standing crop of fish in the pond is approaching carrying capacity (this level is determined by the fish farmer from his experience and is specific to each farm and even to individual ponds), the population should be thinned to allow further growth of the remaining fish. This is usually done at the same time as sampling. If the amount of fish caught in the sample is not sufficient the procedure is repeated a day or two later. If thinning is done for marketing only, a net with larger mesh should be used (35–40 mm and over). This will allow the small fish to escape.

8.4 HARVESTING

Harvesting method is determined by prevailing conditions. Usually the best way involves draining the pond, concentrating the fish in the harvesting sump, and loading them into a tank. If the pond cannot be drained, however, the fish can be harvested by seining. In this way a high percentage, but not all, of the fish can be harvested. Since remaining fish can breed in the pond, this method cannot be used as the exclusive way for harvesting fish. Each pond must be drained at least once a year.

8.4.1 Seining

Seining is a convenient way to harvest certain, but not all, species of fish. Carp, nile tilapia, and hybrid tilapia are generally easy to harvest with seines (except for the Chinese variety of the common carp which avoids the net). *Sarotherodon aureus* bypass seines by burrowing into the mud below the lead line, while silver carp and mullet jump over nets. Seining is done with a net which is spread across the pond. The length of the net should be about one and one-half times the width of the pond, but not more than 150 m (for a pond 100 m wide). A longer net requires expensive changes in its construction; for example, stronger lines, etc. The height of the net depends upon pond depth. At both ends there are wooden brail poles, 5–7 cm in diameter, according to the length of the net and the pulling devices. The central part of the seine is higher, forming a sack in which the fish are caught. It is best to start seining at the deeper end of the pond and pull toward the shallow end. The net is pulled from both sides, usually by two vehicles (especially when the net is long) or by hand. People should be specially assigned to ensure that the net is not raised while being pulled. When the bottom of the pond is muddy, the lead line sinks into the mud and makes the net more difficult to handle. This can cause breakage of the line. In order to avoid this, a modification has been recently introduced, after work carried out by the U.S. Bureau of Commercial Fisheries (Coon et al., 1968). A bunch of 15–25 synthetic binding strings are bound together to form a thick band which is then tied to the bottom line of the net. This skids on the mud and prevents the net from sinking. A similar result is obtained when polyethylene strips about 20 cm in width and about 40 cm in length are attached to the bottom line. The strips are folded over the line so that they stretch about 20 cm (half the length) behind the line. These devices, which prevent sinking of the net, ease the pulling and shorten the time of seining. This is especially important when the net is pulled by hand since the team can be cut down to 4–5 people instead of 8–10.

When the end of the net reaches the embankment at the shallow side of the pond, the net, and the sack in the middle of it, are pulled in. This part of the work is one of the most strenuous in fish culture, especially in large ponds when the net contains large amounts of fish. In some farms a power block, which is operated by a tractor, is used for pulling in the net (Figure 8.9). This too cuts down the number of people required for seining from about eight to four or five and saves energy. Greater ease of manipulation can be achieved by adding a sleevelike bag to the center of the net (Coon et al., 1968). This bag is about 15 m

FIGURE 8.9. Pulling a seine by a power block. Note the string band attached to the lead line.

long, is closed at the bottom but open at the top, and floats. If more silver carp (which usually jump over the net) have to be caught, the bag can be tied and closed from above. It is attached to the net by an opening but can be easily detached. It can be transferred and interchanged among a number of nets. When the net is pulled in, the bag is held open and high enough, by means of a number of iron stakes, to prevent the fish jumping over it. It is best to hold the bag under or near the current of water which should continuously flow into the pond at the time of harvesting.

The adjustment of mesh size to fish size is very important. Nets with large mesh are lighter per unit length and therefore easier to drag. This is especially important when the load of fish in the net is high. The net can also be used for a selective harvesting of fish. Here again, adjustment of the mesh size is important. F. Svirski (personal communication) found the relations between mesh size and fish weight which are summarized in Table 8.1.

When fish are crowded in a seine, some small ones, which could have passed through the net, remain in it.

TABLE 8.1 Maximum Weight of Fish (g) That Can Pass through a Net of Given Mesh Size

Mesh Size (mm)	Carp	Tilapia	Silver Carp	Gray Mullet
20	20	—	30	—
25	40	40	60	—
30	100	80	90	150
35	170	120	—	—
40	270	160	—	—
50	400	270	450	650

It is important to take good care of nets, especially the long seine, because they are quite expensive. Direct sunlight can damage synthetic materials, and rats also cause damage when the net is kept indoors. The best way to store a net is on a drying drum. This is a device made of piping. It can be mobile, constructed on a platform with wheels, with its axis parallel to the direction of movement (Figure 8.10). The net is rolled off when used and rolled on again by hand, or by a motor, after seining. It is stored in a shed with a roof to protect it from the sun, but not in a place so closed as to attract mice.

8.4.2 Harvesting by Draining

When the fish have reached market size and harvesting is to be initiated, feeding is stopped and the pond is drained at a rate which is determined by the capacity of the monk and drainage ditch. During draining and the concentration of the fish in the harvesting sump, as well as during harvesting itself, the supply of oxygen should be maintained by spraying fresh water at a number of points or by aerating with one of the devices used for this purpose (see Section 11.1.3). Without these measures the fish are in danger of weakening and dying. In the harvesting sump the fish are concentrated toward the loading equipment with the help of a net, the length of which is usually not over 30 m (Figure 8.11).

When the amount of fish in a pond is very large, and conditions in the harvesting sump do not allow for holding them safely, seining and draining are used together to harvest the fish. While the pond is being drained it can be seined and some of the fish removed. The remainder are harvested from the harvesting sump in the same way as described above.

FIGURE 8.10. Rolling off a seine net from a net drum into a pond.

FIGURE 8.11. Concentrating fish in a harvesting sump with a seine. Note the water flowing into the sump.

The use of nets can be avoided by employing a concrete harvesting box (see Figure 8.12) which replaces the harvesting sump. This method, described by Pruginin (1959), is based on the tendency of fish to move in the direction of fresh water and against the current, especially when the oxygen concentration in the harvesting sump is deteriorating. For a 4 ha pond, a concrete box 2.5 m high, 10 m long, and 2 m wide is sufficient. For larger ponds (4–6 ha) the box should be longer—up to 15 m. The box should be built in the deepest part of the pond, near the monk, perpendicular to the embankment. The bottom of the box is 80 cm below the bottom of the pond, and the top of the walls are 10 cm above water level. The box has a gate which faces the pond and an outlet to drain it into a nearby drainage ditch. A number of water inlets (three or four) supply fresh water to the pond. After draining about two-thirds of the pond, the gate of the box facing the pond is opened, and fresh water is introduced into the pond through the box at the rate of 20–40 m³/hr. At the same time, draining of the pond is continued. The fish usually follow the current of fresh water and

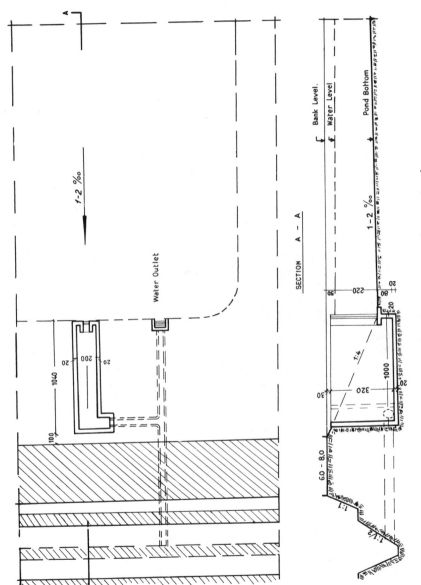

SECTION A – A

FIGURE 8.12. Plan for a concrete harvesting box.

enter the box. When the pond is empty, all the fish are concentrated in the harvesting box and are loaded from there into a tank. The advantage of this box is the saving of manpower and effort in pulling nets, but the main drawback is one of high cost. Their high cost is the major reason these boxes are not widely used on the farms.

A similar metal box has recently been developed. It is made of 3 mm thick iron sheets and is, therefore, less expensive (Doran, 1976). The box is set perpendicular to the embankment. The embankment side of the box is sloping, and the length of the upper part is about 5 m, while the length of the lower part is only 2 m; the width is 1.2 m. The height of the walls above the bottom of the pond is 90 cm, while the depth below the bottom is 40–110 cm, depending on the depth of the drainage ditch and the height of the loading equipment. Overall depth is thus 130–200 cm. The box is adjacent to the monk so that water can be drained throught the box. The method of operating the metal box is similar to that of the concrete box as described above. The metal box is smaller, however, and therefore does not replace the harvesting sump but complements it.

Another method which enables fish harvesting without a net is to let the fish leave the pond with the water through the outlet and then collect the fish in a special box outside the pond (near or in the drainage ditch). The fish are loaded directly from the harvesting box into the transportation tank. This method was practiced in some European countries (Huet, 1970). Some risks are involved, however, in using this method in warm climates. Healthy fish usually tend to swim against the current. They therefore avoid the outlet current and do not pass outside the pond until most of the water has already been drained away. Only when a shallow pool remains in the harvesting sump, causing the fish to concentrate at the outlet, are they swept out by the strong current. Water in the sump warms up rapidly, placing the fish under considerable stress. This may lead to mortalities during harvesting. The risks involved can be reduced considerably by ensuring a flow of cool, fresh water into the harvesting sump and spraying water into the harvesting box outside the pond.

Different species react differently to harvesting stresses. While carp and tilapia are hardy and usually recover rapidly after being transferred to cool fresh water, silver carp and mullet do not, and usually die quickly. This may not make any difference if the fish are harvested for the market and are sold fresh but not alive. If, however, the fish are to be restocked for further rearing, some less stressful harvest technique should be used.

When harvesting outside the pond is anticipated during the physical planning of the farm, the ponds can be constructed to have a common harvesting box for a number of adjacent ponds, thus saving on construction costs.

8.4.3 Loading, Sorting, and Packing

Loading fish from a net into transportation tanks for the market or for transfer to other ponds can be eased considerably if elevators or other means of lifting are used. Most elevators consist of a trolley moving on rails (Figure 8.13). The trolley contains 50–250 kg of fish. The length of the rail is 10–14 m, and the entire device is mobile and mounted on two wheels so that it can be towed from one pond to another (Figure 8.14). For loading, the elevator is installed in the pond so that the lower end is in the harvesting sump near the net (Figure 8.15), and the upper end extends over the embankment to enable the loading of a fish transportation tank on a truck under the head of the elevator. Sometimes a device at the head of the elevator enables distribution of sorted fish into two adjacent tanks which are positioned side by side under the elevator.

A number of new loading devices have been recently developed. The first is a cup conveyor (Doron, 1976). The advantage of the cup conveyor over the elevator is that the fish are not pressed down by their own weight in the trolley while being loaded. The fish are loaded in small amounts, but continuously, so that they can be sorted while on the conveyor belt or on the embankment. Conveyor length can be adjusted between 10 and 15 m according to pond depth and the distance from the harvesting sump to the embankment. The angle of the slope and height of the conveyor above the ground can also be adjusted. A

FIGURE 8.13. Fish elevator lifting fish to a sorting table. Note the three transportation tanks for receiving the different kinds of fish.

FIGURE 8.14. A fish elevator in transport position. It can be towed by a jeep and easily installed in a pond.

conveyor can be installed in a pond within about 10 minutes. The entire system is run hydraulically by the power takeoff of a tractor, and the speed of the belts is regulated by hydraulic valves.

Another new device for loading fish into transport tanks is the vacuum tank (Figure 8.16). This device has four parts: a sealed tank, a vacuum/compression pump, a 15 cm (6 in.) flexible suction hose, and a 15 cm loading hose. The pump is operated by a tractor power takeoff. Operation starts with the closing of the outlet pipe and suction of air out of the tank, creating a vacuum within the tank. The intake of the suction hose is placed in the net where the fish are concentrated. Fish and water are sucked into the tank until the tank is almost full. At this stage, by merely changing a lever, the direction of pumping is reversed and air is compressed into the tank causing the water and fish to be pushed through the loading hose into the transport tank. Just before

FIGURE 8.15. Loading fish onto an elevator's trolley.

FIGURE 8.16. Loading fish by a vacuum tank. Note the sucking and draining hoses.

entering the tank the water is screened off and returned to the pond
while the fish fall into the transport tank (Figure 8.17). Each cycle of
suction and compression takes about 3 minutes and can load 0.5–1.0
ton of fish, depending on their concentration in the net.

Some difficulties were met in "pumping" tilapia. Like other fish,
they tend to swim against the current and are swept into the suction
pipe tail first. When alarmed, however, they spread their fins (especially the dorsal fin), which have very sharp spines. These spines may

FIGURE 8.17. A view of a loading vacuum tank showing the separation of
the fish from the water.

be stuck in the flexible soft plastic walls of the hose and thus block the passage, or the fins may be broken off and the fish then reach the tank damaged and of lower market value. This is remedied by making the sucking hose out of a special material.

While still in the development stage, the vacuum tank has been operated satisfactorily on a commercial scale. The advantages of this method are its speed and the small amount of labor involved.

In polyculture systems the fish have to be sorted by species and sometimes graded into different size groups within species after harvesting. Sorting is usually done manually. The more sensitive fish species, such as silver carp and mullet, if intended for restocking, are often sorted in the pond, directly from the drag net, and loaded into tanks. This shortens the time they are held in the net and reduces stress and mortality. Early sorting is important for silver carp even when the fish are harvested for market since if they are held long in warm water their flesh becomes soft and quickly spoils. Sorting of silver carp from a net is easy since they are usually the largest fish present.

The hardier fish species, such as carp and tilapia, are sorted outside the pond, either on the embankment beside the pond or at the packing center of the farm. In the first case, the sorting is done while the fish are on the conveyor belt, or, more often, it is done on a special sorting table just below the head of the elevator (Figure 8.18). The sorting table consists of a moving belt divided into several lanes which lead to different transport tanks through special shutes. A small portable sorting table is shown in Figure 8.19.

A less common method of sorting is to load all the fish into one

FIGURE 8.18. A sorting table below the head of an elevator allows for dividing fish into three transportation tanks.

FIGURE 8.19. A small portable sorting table.

transportation tank and bring them to a central sorting and packing station as quickly as possible and do the sorting there. Even then, sensitive species are sorted out first in the pond and transported separately.

Grading into different size groups can be done with an adjustable bar grader. Small fish, which usually go for restocking, fall between the bars into one container while the larger fish, usually market size, fall into a separate tank and are packed into boxes and shipped under ice.

In some countries, such as Israel, fish are marketed in two ways: common carp alive, the other species fresh. In order to ease handling and transport for long distances, carp must be "flushed out" before transport (see Section 8.4.4). For this purpose they are held in a storage tank or pond for at least 8 hours. These premarketing storage tanks can be constructed of concrete to ease loading into the transportation tanks (see Figures 8.20 and 8.21). They are divided into a number of compartments, each holding one transport load of fish (3–5 tons). Grading is often done over these tanks so that the small fish fall into one compartment and are taken from there for restocking, while market size fish fall into another compartment and are sent to market. The tanks are equipped with an aeration system and water flow to ensure the supply of oxygen and to maintain the fish in good condition until they are restocked or marketed.

Fish that are marketed not alive but fresh are transferred to the packing system. Since in many regions water and air temperatures are high during most of the year, spoilage of fish is quite rapid and freshness must be maintained by rapid cooling after harvesting. This is especially important with fish such as the silver carp, which spoil quicker than other species. For rapid cooling the fish can be introduced into a cold-water tank, the temperature of which is maintained at

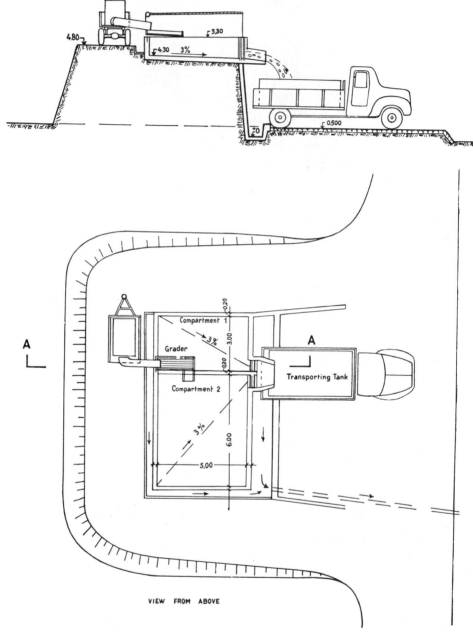

FIGURE 8.20. Plan of a concrete storage tank.

FIGURE 8.21. Loading fish from a concrete storage tank into a transportation tank. (Courtesy of S. Rothbard, Gan-Shmuel, Israel.)

1–2°C. Cold shock kills the fish immediately and permits a more convenient and efficient way of working and packing. Moreover it was found that the shelf life of fish killed in this manner is longer and their external appearance better. They therefore demand higher prices in some markets. Fish are held in the cold-water tank for about half an hour to cool the entire body. They are then graded and unmarketable fish removed. Marketable fish are packed in boxes, covered by a plastic sheet, and covered with flakes of ice. If they are not marketed immediately, they should be stored in a cooler.

8.4.4 Fish Transportation

In many cases fish have to be transported alive—on the farm itself, between farms, or to the market. Such transportation is done in tanks constructed in most cases from sheet iron. These tanks may have various volumes, according to their purpose. See Figures 8.22 and 8.23. For transporting fish within a farm, the volume is between 1 and 8 m³. For transporting fish to market, the tank contains about 12–16 m³.

FIGURE 8.22. A tank for transporting fish to the market being loaded on a truck. Note the air blower unit on top of the front part of the tank. (Courtesy of Ray Coleman, Ma'agan Michael, Israel.)

Such a tank can transport about 5 tons of fish at a fish/water ratio of 1:2 over about 5 hours. These larger tanks are usually loaded on trucks for transport purposes.

Loading and transport exert considerable stress on the fish. Factors affecting condition and survival of the fish during transportation include:

Physiological State of the Fish. The fish should be healthy and should not be fed before transportation. Fish with full stomachs require larger amounts of oxygen for digestion and may perish during transportation to market.

FIGURE 8.23. Tank for transporting fry. Oxygen is supplied from two cylinders, one on each side of the tank.

Oxygen supply. This is usually supplied by aeration using an air blower and a perforated pipe system at the bottom of the tank. The blower is operated by a small gasoline engine or the tractor's power takeoff. When no blower is available, or when the fish are sensitive to agitation, as in the case of small fry (especially mullet fry), oxygen is supplied from a compressed oxygen bottle through special diffusers made of perforated rubber tubing or special plastic tubing. These release oxygen in small bubbles which do not harm the fry.

Water Temperature. The optimal temperature for fish transport (except tilapia) is 10–12°C. At this temperature respiration is reduced, and the ratio between fish and water can be increased while maintaining the fish in good condition for quite a long period. At a fish/water ratio of 1 : 2 it is safe to transport fish over 24 hours. The temperature can be reduced by putting ice into the tank. Care should be taken that the ice does not contain any chemicals that may poison the fish.

Ratio between Fish and Water (Weight/Volume). For short distances, such as on the farm, the ratio of fish to water may be increased up to 1 : 1. For longer distances, however, when the transportation may take 3–8 hours, the fish/water ratio should be reduced to 1 : 1.5 or 1 : 2. This will depend on the water temperature. If the temperature cannot be reduced, the smaller ratio is recommended.

Tilapia and mullet are more sensitive to stress caused during transport. They may lose scales, due to friction among themselves or due to handling, and die either during transport or, more often, through infection thereafter. Care should be taken, therefore, not to overcrowd them and to increase the fish/water ratio to 1 : 3 or even 1 : 5 for mullet fry and not to overagitate the water.

8.5 WEED CONTROL

There are several ways in which weeds may affect production of fish in ponds. Dense vegetation restricts the movements of fish and limits space and the availability of natural food. Floating plants such as the yellow waterlily (*Nuphar luteum*) or duckweed (*Lemna* spp.) sometimes completely cover pond surfaces and prevent light penetration into the water. This can reduce photosynthesis and oxygen production by algae and may lead to oxygen depletion and subsequent fish mortality by anoxia. Weeds assimilate a large proportion of the soluble nutrients in the pond and thereby affect plankton production and the production of other natural food organisms. Some weeds, such as the reed (*Phragmites communis*) and cattail (*Typha* spp.), trap large amounts of silt and organic detritus around their roots and stems (Gaudet, 1974). Within a few years such plants can cause a considerable portion of infested ponds to fill up. In addition, dense vegetation in ponds creates a physical obstacle to seining for harvest and sampling. Pruginin (1968b) cites cases in which ponds heavily infested with submerged weeds such as *Potamogeton* sp. and hornwort (*Ceratophyllum demersum*) yield only 600–700 kg/ha of fish. Production increased to 2000 kg/ha after the weeds were eradicated by the application of sodium arsenite.

Plant species which are specific in terms of water depth form a characteristic vegetation zone. There is a certain amount of overlapping at the margins of the zones, but each zone may be distinguished by its dominant species. Three such zones can be identified in fishponds:

The Embankment Zone.   Plants that grow in this zone do not invade the water, but they do need a high level of soil humidity for their roots. The species will differ in different climates but the most common, having wide distribution, are the hairy willow herb (*Epilobium hirsutum* L.), *Inula viscosa, Lycopus europeus, Rubus saguineus,* and *Apium graveoleus.*

Emergent Vegetation Zone.   Emergent aquatic plants of this zone grow in water depths not exceeding 50 cm. These weeds can germinate in wet soil and send shoots through the water column into the sediments. Their roots grow in submerged or damp soil, but most of the stem, leaves, and flowers grow above water level. The more common species are reed (*P. communis*), cattail (*Typha latifolia* and *T. augustata*), and rush (*Juncus* sp.)

Aquatic Zone.   This zone, which covers depths of 50–200 cm, has vegetation consisting of rooted submerged plants such as the hornwort (*Ceratophyllum* sp.), water milfoil (*Myriophyllum* sp), and pondweed (*Potamogeton* sp.); and rooted plants with foliage floating on the surface of the water such as the broad-leaved pondweed (*Potamogeton natans*), amphibious bistort (*Polygonum amphibium*), and free-floating plants such as duckweed (*Lemna* sp.). A considerable nuisance is created by filamentous algae such as *Spirogyra, Cladophora, Pithophora,* and others and at times also by planktonic microalgae. Blue-greens such as *Microcystis* that float on the water may form a scum which can adversely affect the oxygen regime in water.

The weeds of the embankment zone are the least noxious, and often plants of this zone, such as *Lippia nodiflora,* are useful in consolidating the soil of embankments and help prevent erosion. They may, however, prevent free passage on the embankment and thus disturb work on the farm. Weeds of the emergent and aquatic (submerged vegetation) zones are the most noxious. The extent of the former zone may, however, be reduced by constructing ponds so that the shallowest part will be at least 80 cm deep.

Three methods of controlling vegetation are practiced in fish ponds: (1) mechanical, (2) chemical, and (3) biological.

8.5.1 Mechanical Control

Mechanical means include mowing, dredging, and burning are used in large numbers of fish farms. The different techniques and apparatus used are numerous and varied. These can be either simple instruments or mechanical weed cutters. For small ponds a scythe is generally used, but for larger ponds mechanical aquatic weed cutters have been developed, mainly in Europe. Robson (1974) gives details on some of these machines. The main drawback of all these methods is their low efficiency in keeping ponds clear of vegetation, for moving and burning have to be repeated several times a year. Also, owing to increasing labor costs and the shortage of manpower in many countries, they have become uneconomical.

Blackburn (1974) points out that in the United States submerged weeds may be controlled with herbicides at one-fourth to one-third the cost of mechanical means. In Israel, the number of days for manual weed control per hectare of pond per annum worked out at 30–50, which was 30–50% of the total working days involved in fish farming. The introduction of herbicides in 1957 eventually reduced the time involved in weed control to 5–10 days/ha/yr. It was also found that the total cost of herbicidal control was as low as 20% of that of manual labor. It is obvious that where labor is not expensive the situation may be different, but this should be carefully calculated.

8.5.2 Chemical Control

The ready supply of herbicides on the market and the uncomplicated equipment for their application have led to their widespread use for controlling pond weeds. There are three methods for the use of herbicides for this purpose: soil sterilization, foliage treatment, and application in pond water.

Soil Sterilization

Herbicides used to cause sterility of the soil are applied in the embankment zone. The expense involved in this method makes it impractical for most large ponds. It may, however, be used successfully in small ponds, limited areas of embankment, or for spot treatment of relatively clean areas. Among the products available, CMU [3-(p-chlorophenyl)-1,1 dimethyl urea] is intended for sterilization of mixed vegetation areas of dicotyledons and monocotyledons, but not for deep-rooted perennial plants. In ponds, however, CMU was found to be too expensive. Simazine [2-chloro-4,6 bis (ethylamine)-triazine] is much more efficient in eradicating seasonal weeds on the embankment and preventing their germination. Spraying the soil should be done immediately after rain has fallen, and the chemical should be used at the rate of 1.5 kg per hectare of embankment. Simazine is absorbed by the roots of the plants. It is also absorbed well by algae and submerged plants and is therefore used to eradicate them. Simazine has no effect when sprayed on the leaves of higher aquatic plants. Products containing sodium chlorate are suitable for controlling deep-rooted vegetation, such as reeds (*P. communis*), *Prosopis furcata,* etc. As a result of their high solubility these chemicals rapidly penetrate to a depth which CMU only reaches after a very long time. Deep-rooted weeds are, however, eradicated more efficiently (and less expensively) with Dalapon.

Foliage Treatment

Foliage treatment has only a negligible effect on the soil and is used mainly in the emergent vegetation zone. The herbicides used can be subdivided according to their action into broad spectrum herbicides which destroy most plants and selective herbicides which kill only certain plants. Although most of the broad spectrum herbicides do not penetrate the plant material to great depth, they are effective on contact with plant tissues. Control of perennials with contact herbicides requires thorough and continuous treatment whenever growth occurs. Their use in ponds is therefore limited.

Selective herbicides are used in the control of specific plants or groups of plants, leaving others relatively unaffected. The most widely used of these are 2,4-D and 2,4,5-T, which have a marked selective action on dicotyledonous or broadleaf plants (although in some cases they do affect a number of monocotyledons), while Dalapon is toxic to many grasses and other monocotyledons, but usually does not have a serious effect on broadleaf plants.

Some of the more widely used herbicides, the plants that can be controlled by them when sprayed on their leaves, and the effective concentrations are summarized here.

2,4-D (2,4-Dichlorophenoxyacetate). Four types of chemical preparations containing 2,4-D are marketed: (1) as an acid, (2) sodium salts, (3) amine salts, and (4) esters. The two more commonly used are the amines and esters.

The 2,4-D (amine) provides effective control against broadleafs such as *Inula viscosa, Prosopis furcatus, Myriophyllum spicatum,* and water hyacinth (*Eichhornia crassiples*). It is sprayed on the foliage as a 1% aqueous solution at a rate of 2–4.5 kg/ha.

2,4-D (esters) control cattail (*Typha angustata*), *Scirpus litoralis, Cyperus* sp., arrowhead (*Sagittaria*), and waterlily (*Nymphaea*) by foliar spray at the rate of 4.5–9 kg/ha.

2,4,5-T (2,4,5-Trichlorophenoxyacetate). This compound is more suitable for control of woody vegetation and shrubs such as brambles (*Rubus* sp.), *Conyza diocoridis,* and smartweed (*Polygonum* sp.). It is applied as a foliar spray with a 0.5–0.7% aqueous solution (or in diesel oil carrier with 0.25% of emulsifying agent) at a rate of 4.5–9.0 kg/ha.

Dalapon (2,2-Dichloropropionic Acid). This herbicide has proved to be most successful for controlling some of the most noxious weeds in fishponds, such as the reed (*Phragmites communis*), cattail (*Typha* spp.), and maidencane (*Panicum* sp.). It is also applied as an aqueous solution of 0.7%, or a 5% solution in a diesel oil with an emulsifying agent. The rate is 6–11.5 kg/ha. Dalapon has been found safe to fish at concentrations of up to 3000 ppm (Blackburn, 1968) and degrades quickly by microbial activity in soil and water (but not in plants). The optimal height of reeds and cattails to be sprayed by Dalapon is about 1 m.

Diquat [6,7 Dihidrodipyrido (1,2 a : 2', 1'-c) Pyrazidiinium Salt]. Diquat has received considerable attention as an aquatic herbicide since the 1960s. It is safe for fish at concentrations much greater than those recommended for aquatic weed control and has no adverse effects on plankton. The herbicide is highly water soluble and easy to apply to ponds. It can be used for both emergent plants such as the water hyacinth (by a foliar spray of 1.3 kg/ha) and submerged plants. However, since it is absorbed very rapidly by suspended soil or organic particles, its potency is reduced to a considerable extent when these conditions exist. This rules out its use in warm-water fish ponds, especially those with carp. In such ponds treatment should be provided before stocking.

Application in Pond Water

Application to pond water is done to control submergent plants and algae. Naturally, with this method of application the problem of herbicide toxicity to fish becomes much more acute, and a wide safety margin should be considered when using these herbicides. Special precautions should also be taken in eradication of filamentous algae and submerged weeds due to the indirect effects of the treatment on fish. Some of the herbicides (and algacides) such as Simazine and copper sulfate have a long-term effect on the production of natural food, either by their nonselective toxicity and persistence in the pond water, or by accumulation on the bottom when used repeatedly. The dead weeds start to decompose a few days after application of the herbicides, taking up large amounts of oxygen, which may also cause the death of fish due to anoxia. Special watchfulness is therefore required here and, if necessary, aeration or changes of water should be employed. The herbicides used in ponds are:

Copper Sulfate.   Copper sulfate ($CuSO_4$) was the first and probably is still the most economical chemical used in algae control. Many algal species can be controlled with a concentration of 1–3 ppm copper sulfate pentahydrate (CSP). However, some species such as *Pitophora* spp. are more resistant and require higher concentrations of CSP. Most of the emergent plants are also resistant to CSP.

The main drawbacks in the use of CSP are that it is toxic to fish and many invertebrates and the safety margin is quite narrow. Copper sulfate should be used only if necessary, therefore, and with great care.

Kessler (1960) controlled blue-green *Microcystis* blooms in ponds with copper sulphate by spraying the scum which is usually concentrated by the wind at the leeward side of the pond. He used a 3% solution of CSP. This gradually controlled these algae without causing harm to the fish. Note that copper sulfate may corrode iron, so sprayers must therefore be made of resistant material.

Sodium Arsenite.   Sodium arsenite ($NaAsO_2$) is another inexpensive and effective herbicide for submerged plants. The use of this chemical is restricted due to its toxicity to mammals and aquatic fauna. It is effective in controlling *Potamogeton nodosus, Elodea, Myriophyllum,* and unbranched filamentous algae such as *Hydrodictyon* and *Cladophora* when used at a concentration of 4–5 ppm active ingredient. The commercial product usually contains about 85% active ingredient. Most fish will withstand concentrations higher than 4–5 ppm. Because of a possible buildup of arsenic in bottom muds and fish, and because of the danger of toxicity to personnel (0.02 g will kill a man), the use of sodium arsenite is discouraged.

Diorex. Diorex is a relatively new herbicide for filamentous algae and submerged plants. It is not as toxic to fish as the previous ones, but it can, of course, cause indirect damage by anoxia. Diorex is applied at the rate of 1–2 ppm in the water.

As pointed out above, the use of herbicides saves labor and expense as compared to manual weeding. The major expenditure on herbicides is for the first application. Later applications are done only upon the appearance of isolated patches of weeds. These require much less spray and work.

In the shortterm, herbicides can be effective in eradicating pond weeds. Not much is known, however, about the long-term effects of herbicides on ponds and fish (such as herbicide accumulation in fish flesh, for example). It is recommended therefore that herbicides be used as sparingly as possible. Remember that proper management of fishponds (depth of water, planting of embankments with grass, etc.) and biological control can reduce weeds and so reduce the need for herbicides.

8.5.3 Biological Control

Weeds on embankments can be controlled either by grazing or by preventing their emergence with a cover of grass. Grazing by cattle is efficient only for a short time. Cattle graze only on specific weeds such as *Phragmites communis* and *Panicum repens.* They do not touch others such as *Inula viscosa* and *Typha angustata.* As a result, weed species composition changes within a few years. Pond embankments which were infested mainly with *Phragmites* become covered with only *Inula* and *Typha.*

The fur animal nutria (*Myocaster coypus*) was introduced into fishponds in Israel for the purpose of producing furs and meat, on the one hand, and clearing the ponds of weeds on the other. Experiments conducted showed that the nutria are efficient in controlling *Typha, Phragmites,* and *Polygonum* sp. However, the animal does not eat *Inula viscosa, Juncus bufonius,* and *Apium graveolens.* In time it became evident that there is no economic justification for holding nutria in ponds for their fur and meat. It is necessary to fence ponds to protect nutria from wild animals and to prevent them from escaping. This involves a large investment. Nutria which escape from the fenced area may, if their population becomes uncontrolled, cause severe damage by burrowing into embankments or by making holes in the sluice boards of the monk, causing a drain-off of pond water and the loss of large amounts of fish. On the other hand, in some places where nutria have natural enemies, their population is balanced and effectively reduces weed density (especially cattail). Since the price of nutria fur on the world market is low, it produces little or no profit. The use of nutria solely for biological control of weeds in ponds should, therefore, be very carefully considered.

Planting Kikuyu grass (*Prisatum* sp.) as well as other grasses along the banks above the water level of a newly constructed pond creates a dense cover which prevents the colonization of weeds and the erosion of the embankment itself. Such grass should be planted immediately after the pond is filled.

Weeds in the emergent and submergent vegetation zones can be biologically controlled by herbivorous fish. Principal species are the grass carp (*Ctenopharyngodon idella*), herbivorous tilapia (*Tilapia rendalii* (=*melanopleura*), and *Puntius javanicus*. These fish do not feed on all weeds, and therefore plants which are not eaten can cause problems. The population of grass carp required to check the growth of weeds is 100–200 fish/ha.

Another kind of weed control by fish involves submerged weeds such as *Ceratophyllum* and *Myriophyllum* and filamentous algae (*Spirogyra* sp.). An intensive growth of these weeds usually occurs in spring. Carp do not feed on these weeds but burrow into the bottom mud and prevent the support of weeds by muddying the water and reducing photosynthesis. In order to do this the fish must be stocked at densities of about 2500 fish/ha at weights averaging over 250 g. If the ponds are heavily infested by plants they could be cleared within 20 days by turning them into storage ponds for carp of average weight of 500 g or more and densities of 8000 fish/ha.

When mullet fry are stored or nursed in ponds after collection from estuaries in the autumn, and when tilapia fry are nursed, a dense filamentous algae bloom often develops. This can cause difficulties in harvesting fry since they can become entangled leading to heavy losses. The growth of algae may be prevented by stocking carp of 500–600 g each at a density of 250–300/ha. In order to avoid "wild" spawning of carp, and to resolve the problem of sorting fry from other species nursed in the ponds, only male carp are used.

Another biological control method for submerged weeds and filamentous algae such as *Najas, Potamogeton* spp., and *Chara* spp., suggested by Smith and Swingle (1941), is to heavily fertilize ponds with chemical fertilizers. The effect of the fertilizer is not direct. It induces heavy growth of phytoplankton which, by shading submerged plants, hinders their growth. From experience it seems that this method can be used as a preventive measure immediately after filling a pond and before the appearance of weeds. The weeds will not grow in phytoplankton-rich water because of low light penetration. Where weeds are already established, however, this method is not practical.

8.6 KEEPING RECORDS

All activities on the farm should be recorded. This includes both those things initiated by the farmer himself such as stocking, feeding, sample weighing, fertilizing, and harvesting; and natural phenomena such as diseases, fish kills, anoxia, etc. The records provide information on the current state of the ponds during the growing period, and regular recording of information for a number of years supplies information necessary for determining natural pond productivity, the effect of different management methods on yields, the best species combinations (see Section 7.5.1), and the combinations' effect on profitability. Farmers in Israel have developed an information card (Figure 8.24) which has proved convenient. It is used by most farmers in the country. The use of a uniform card on all farms within a region enables a comparative analysis between and among farms using uniform parameters.

Data on sample weighings and feed conversion ratios let one examine the situation in periods between sample weighings and pinpoint the occurrences of lower growth rates and/or low feed utilization. Data on standing crops, average weight of fish, and other factors can help in determining the reasons for such occurrences. In order to make the record as useful as possible, all relevant data must be recorded on the card.

A yearly sum-up sheet for an entire fish farm is shown in Figure 8.25.

9   Fertilization and Manuring

9.1 THE NATURAL FOOD WEB

From our earlier discussions (see Section 7.1) we have already learned the importance of natural food for fish growth and pond yield. Natural food, which is produced in ponds at almost no cost, replaces costly supplementary feed. Since natural food is rich in protein, vitamins, and other growth factors, simple supplementary feeds containing relatively low levels of protein and vitamins can be used. This affects the economy of culture because such feeds (e.g., cereal grains) are much less costly than the complete diets used in running water systems where there is no natural food. Except for special cases where market prices for warm-water fish are relatively high, such as in Japan or for catfish in the United States, it is doubtful whether the profitable culture of fish such as carp, tilapia, and Chinese carps, which get lower market prices, would have been possible without natural food. With only "artificial" feed and no natural food, the cost of nutritionally balanced complete feed would be so high that there would be no profit.

It is worthwhile, therefore, to increase the production of natural food in the pond as much as possible. This will allow an increase in fish density with no appreciable decrease in individual growth rate, and will thus increase total yield. Alternatively, less supplementary feed can be provided with a resulting decrease in the apparent feed conversion ratio. Increasing the production of natural food is achieved by introducing chemical fertilizers and/or organic manures. In order to use these efficiently, one should first understand their effect on the pond food web.

Under natural conditions, each link of the food chain depends on its supply of food from lower trophic levels. Thus, if the main food of adult carp is chironomid larvae which develop in the pond bottom, the development of chironomids will depend on detritus, bacteria, and other small organisms developing on this detritus. The amount of detritus depends, of course, on the production of organic matter by algae (assuming that very little organic matter is introduced into the pond from the outside). Fish, such as bighead carp, feed mainly on zooplankton, and the abundance of their food will in turn depend on the algae which the crustaceans consume. Similar food chains can be described for all fish in a pond. The first link will always be algae, the main primary producers present, which convert mineral nutrients into organic matter.

In reality, food chains are much more complicated than described

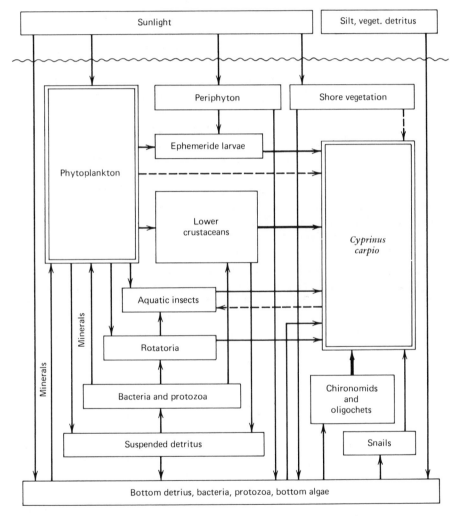

FIGURE 9.1. Schematic representation of the food web ending in common carp. The size of the rectangle does not represent the importance or the volume of the group in the food web.

above. The food of a fish never consists of one specific item. There is a degree of preference among a number of natural food items when natural food is abundant and selection is possible. However, when natural food is difficult to obtain, fish feed on what is available rather than on what is preferred. Thus, fish depend on a number of natural food items each of which, in turn, is dependent upon various other food sources. The food chain can therefore be better described as a *food web*. Figure 9.1 presents an illustration of the food web of common carp similar to that of Vaas and Vaas-van Oven (1959). The basis of the web remains phytoplankton. This is the only autotrophic (organic matter producing) component of the web; all other components are heterotrophs—they consume organic matter.

It is logical to assume, as was done in Germany at the turn of this century, that by stimulating primary production one can stimulate the production of all other trophic levels and therefore affect fish yield. This can be done by chemical fertilizers which supply the minerals required for production of organic matter through photosynthesis. Indeed, early experiments carried out in Germany and later in many other places have shown that a considerable increase in fish yields can be achieved by chemical fertilization. Fertilizer doses, frequency and methods of application, as well as some limitations of chemical fertilization, are discussed in the next section.

Both minerals and light are required for photosynthesis by phytoplankton. It was found that when minerals are present in sufficient amounts, and when, as a result, the density of phytoplankton in pond water increases, light penetration into the water decreases. This limits primary production by phytoplankton in the lower levels. Figure 9.2 shows primary production at different water depths in ponds with two levels of fertilization as compared to production of a pond which does not receive fertilizers. While fertilizers considerably stimulate primary production, doubling the amount of fertilizers does not produce higher primary production because of the autoshading effect. This is also reflected in carp yields from the ponds. The larger amounts of fertilizers result in only a slight increase in carp yield.

One way to overcome light limitation is by stimulating the production of heterotrophic links of the food chain with the addition of readily decomposable organic matter. Bacteria which develop on organic matter serve as food for many other links in the food web. Protozoa, for example, are readily eaten by many organisms, including fish.

Organic manure should, however, be applied with care. Rapid decomposition of organic matter, and the increased bacterial population associated with it, may result in complete oxygen depletion and a fish kill.

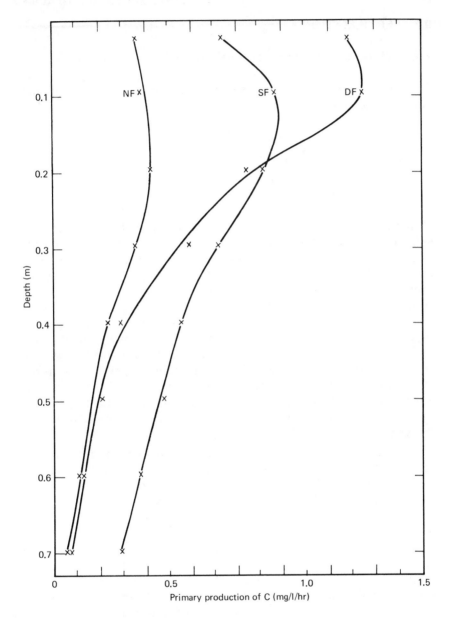

FIGURE 9.2. Primary production–depth curves of ponds receiving different levels of fertilization. NF, No fertilization; SF, "standard" fertilization; DF, double fertilization.

9.2 FERTILIZATION

9.2.1 Need for and Time of Fertilization

Three main factors affect the need for fertilization, the nutrients to be applied, and their amounts:

1 Demand for natural food by the fish.

2 Nutrient requirements of phytoplankton.

3 Availability of nutrients in the water.

During winter (in Israel, from the end of November to the end of March), the temperature of pond water is relatively low (10–18°C). The fish, being poikilotherms, conform to ambient temperature. Metabolism and growth of warm-water fish are much reduced in low temperatures. Food is only required in limited amounts for basic maintenance. Thus, the food requirement in this season is very low. The production of plankton, though somewhat lower than in summer, is not as low as may be expected from the decrease in temperature. Some algae and zooplankton species, more adapted to the lower winter temperatures, replace the summer species, and production remains fairly high. The effect of fertilizers under these conditions is limited, mainly because of the limited use of the natural food by fish. In Israel, therefore, ponds are fertilized only in summer. This may be different, of course, in tropical regions, where both fish and plankton grow well throughout the year.

If the nutrients required by algae are supplied through a natural source within the pond there is, of course, no need for supplementation with fertilizers; or, this supplementation becomes less critical for the development of the phytoplankton. In shallow fishponds at the Fish and Aquaculture Research Station at Dor, appreciable amounts of water are added in summer in order to make up for losses by evaporation and seepage. Hepher (1962b) has shown that since the water contains a relatively high concentration of nitrogen (about 2 mg/l), the effect of fertilizing with nitrogen fertilizers is negligible. The same kind of fertilization had an appreciable effect in spring and autumn when less water was added to the pond (Figure 9.3).

Another way of supplying nutrients to the water is through organic matter which, upon decomposition, releases nutrients in soluble form. It is logical to assume that with intensification of fish culture, when the density of fish in a pond is high and they are fed large amounts of protein-rich feed, both nitrogen and phosphorus are being

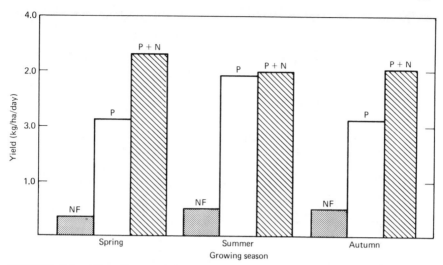

FIGURE 9.3. The effect on fish yield of fertilizing with phosphate (P) alone at a rate of 60 kg/ha every 2 weeks, or with phosphate and nitrogen (P + N) at a rate of 60 kg/ha every two weeks, with no supplementary feed added, during three growing seasons (of about 80 days each).

supplied in higher amounts. One should remember that at least 20%, and usually much more, of the nitrogen ingested by a fish is excreted into the water. Much of the phosphorus in the feed has a similar fate. No experiments have been carried out on the effect of fertilization in highly intensified fish culture systems.

The supply of nutrients by organic matter is further augmented when a pond is manured. As explained previously, organic manure is added in order to stimulate heterotrophs and to bypass the light limitation on photosynthesis. At the same time, however, organic matter also releases large amounts of mineral nutrients through its decomposition by bacteria. It is not clear whether applying chemical fertilizers is still required under these conditions. Recent experiments carred out by A. Ben-Ari (private communication), in the Upper Galilee, have shown that fertilizing manured ponds with phosphates appreciably increases fish yield. However, this should be studied further under various conditions.

In some cases fertilizers are applied for purposes other than to increase natural food production. Three such cases are:

1 Superphosphate is sometimes used to remedy oxygen depletions. Though this practice is controversial, it seems that it may be beneficial in a case of oxygen deficiency caused by the sudden death of algae due to senescence (see Section 11.1.3).

2   Ammonium sulfate or liquid ammonia is applied to control the toxin-producing algae, *Prymnesium parvum* (see Section 11.2).

3   In some cases nitrogen fertilizers are applied in order to control the earthy taste and odor in fish (see Section 11.3).

9.2.2   Kinds of Fertilizers

Most of the nutrients required by phytoplankton are supplied by the water since they are present in solution in adequate amounts. Hepher

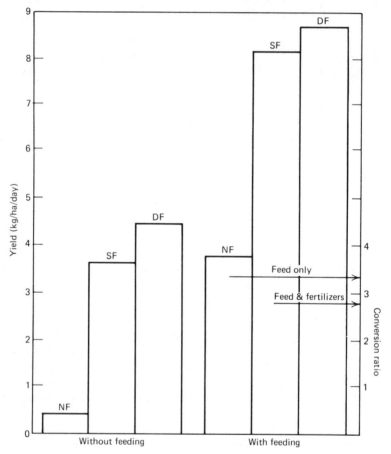

FIGURE 9.4.   The effect of fertilization at different rates on fish yield with and without supplementary feed. Feed amounts were equal for fertilized and unfertilized ponds. The conversion rate is given for feed and fertilizers together without considering their different prices. NF, No fertilization; SF, "standard" fertilization; DF, double fertilization.

(1962b) has found that in semi-intensive culture conditions, where 1200 carp/ha were stocked with no feeding, or 2000 carp/ha were fed cereal grains, both phosphorus and nitrogen were lacking, and the addition of these minerals by fertilization increased yields and decreased feed conversion rates considerably. Figure 9.4 presents a summary of the results obtained during 10 years of experiments.

Since the main rock formations and thus also the soil and the water in Israel contain relatively high concentrations of potassium (Table 9.1), it was assumed that this mineral was available in sufficient amounts. Determinations of potassium concentrations in the pond water at the Fish and Aquaculture Research Station at Dor fluctuated only between 9.0 and 10.0 mg/l. It seems, therefore, that any uptake of potassium by phytoplankton, or loss of potassium from the water by any other means, is balanced by the amounts added with the water. No experiments have been conducted in regions such as the Upper Galilee, where natural potassium concentrations are lower (Table 9.1).

Only dissolved minerals can be utilized by algae. Nutrients may be present in large amounts in the bottom soil, or even in suspended insoluble form in the water, but they will still be unavailable to the phytoplankton until dissolved. Obviously, added fertilizers must also be dissolved in water. This is not a problem with nitrogen fertilizers, all of which are water soluble and can, therefore, be applied directly to fishponds. The most common nitrogen fertilizer is ammonium sulphate which contains about 21% N. Urea is also sometimes used. This is an organic compound, and the nitrogen in it cannot be absorbed by algae. Urea, however, is easily decomposed into an inorganic form. Urea is concentrated, containing about 46% nitrogen. In some places liquid ammonia is cheaper than other nitrogenous fertilizers. It is a solution of NH_3 in water which is in equilibrium with NH_4OH. It contains about

TABLE 9.1 Potassium Concentrations (mg/l) in Waters from Different Regions of Israel[a]

Region	No. of Determinations	Max.	Min.	Avg.
Coastal Plain	26	23.8	0.2	6.0
Zevulun Valley	7	18.0	6.4	13.5
Beit Shean Valley	42	29.7	0.0	8.0
Jordan Valley	39	24.0	6.0	10.0
Upper Galilee	32	3.5	0.0	0.9

[a] From Hepher (1962b).

20.5% N by volume and requires special methods of application (see Section 9.2.4).

Most phosphorus compounds are not soluble in water. Some of these such as bone phosphate and dicalcium phosphate, though sometimes used for fertilizing soils of low pH, are not suitable for fishponds since they are not available to the phytoplankton. The most important phosphate fertilizers for ponds are the superphosphates. Single superphosphate is made by treating rock phosphate with sulphuric acid. The resulting product is a mixture of monocalcium phosphate and calcium sulfate (gypsum). It contains 16–18% P_2O_5 (about 7% P). A more concentrated superphosphate is called "double," "triple," or "treble" superphosphate. This fertilizer contains 40–50% P_2O_5 (about 17–21% P). Any of these can be used for pond fertilization by taking into account their relative phosphate contents. Experiments have been carried out with ammonium phosphate ("monophos") and phosphoric acid (Hepher, 1958). They seem to be retained for a somewhat longer period in a solution in water than superphosphate. This may be due to the slightly lower pH associated with them, to the absence of calcium in the compounds, and to their high degree of solubility and thus lesser contact with the bottom mud. There was no significant difference, however, in fish yield through the use of these fertilizers compared with yield through the use of superphosphate.

9.2.3   Fertilization Rates

In addition to the nutrient requirements and their availability, the rate of fertilization is also affected by chemical reactions in pond water. Quite a stable equilibrium exists between the phosphorus dissolved in pond water and that contained in the solids, mainly the pond's bottom soil (Hepher, 1966). A similar equilibrium exists between the nitrogen concentration in water and that in the atmosphere (Hepher, 1959b). The application of either phosphate or nitrogen in soluble form disturbs this equilibrium, which tends to be rapidly reestablished. In the case of phosphate fertilizers, the excess phosphate precipitates in the form of insoluble compounds; and in the case of excess nitrogen, it is released as gaseous ammonia into the atmosphere. The higher the concentration of soluble nutrients above equilibrium level, the faster are these processes of stabilization. It has been found (Hepher, 1963, 1967) that from a practical point of view there is no biological or economic justification for adding doses higher than 0.5 mg of P per liter or 1.4 mg of N per liter. Any amount above these doses is fixed or released so quickly that it has little benefit. For a pond 0.8–1.0 m deep (= 8000–10,000 m^3 of water per hectare), this equals about 60 kg/ha single superphosphate and about 60 kg/ha ammonium sulfate or liquid ammonia. Thus, this amount has become the "standard" dose for fertilizing semi-intensive fish culture systems with densities of 2000–3000 fish/ha.

One school of thought claims that since there is such a stable equilibrium between phosphates in water and bottom soil, once the phosphates in the bottom have reached a high concentration, there is no need for further application of fertilizers. The assumption is that any uptake by the algae will be made up by the transfer of phosphates from an insoluble to a soluble form so that the equilibrium is reestablished. While in principle this is true, the limiting factor in such systems is the rate of transfer from the solid to soluble form in water. Experiments on primary production in ponds (Hepher, 1962a) have shown that the average production of phytoplankton in fertilized ponds during the summer in Israel is equivalent to a carbon uptake of about 5 $g/m^2/day$. From Figure 9.2 it is clear that this is the maximum production that can be attained until temperature and light become limiting. From ratios among carbon, phosphorus, and nitrogen in phytoplankton, the rate of phosphorus and nitrogen requirements can be calculated, using the above carbon production value. If we take the values given by Fleming (1940) and Strickland (1960) for the relationships among carbon, nitrogen, and phosphorus in phytoplankton (1 : 0.18 : 0.024, respectively), the requirements for nitrogen and phosphorus will amount to 0.9 and 0.12 $g/m^2/day$, respectively. Or, for a pond of 1 m depth (1000 l/m^2), the need for nitrogen and phosphorus will be 0.9 and 0.12 mg/l/day, respectively. Hepher (1966) has shown that this amount is much higher than that which could be transferred from the bottom, and thus production at this level depends on a supply of nutrients from fertilizers. The same seems to be true in ponds in Europe (Hepher, 1966). Indeed, experiments have shown that fertilizing only once or twice at the beginning of the growing season (instead of frequent application during the season), even when high doses are applied, results in lower yields.

Hepher (1963) has studied frequency of fertilizer application and has compared the application of the "standard" dose once every week to its application once every 2 weeks. When the ponds were stocked with carp (1200 carp/ha) and no supplementary feed was added, the difference in fish yield was large in the first two years, but decreased and almost disappeared in later years. After the third year the difference in favor of the higher total dose of fertilizers was statistically insignificant. Average yields for the two treatments during 8 years of the experiment are presented in Figure 9.4. The marginal amount of fertilizer used for each kilogram of fish obtained in the ponds fertilized every 2 weeks over the yield obtained in unfertilized ponds was 3.3 kg. However, 11.7 kg of fertilizer had to be used for every additional kilogram of fish obtained in ponds fertilized every week. In ponds stocked with carp (2000 fish/ha) fed cereal grains, the difference in favor of ponds fertilized every week was even smaller, and, 7.9–16.4 kg of fertilizer had to be used for every kilogram of fish obtained over that in ponds fertilized every 2 weeks. This seems to have been due to light limitation on primary production, as explained in the preceding section.

Yashouv and Halevy (1972) found that when a pond is stocked with silver carp in polyculture with common carp and tilapia, differences in common carp and tilapia yields between ponds receiving "standard" fertilization every week as opposed to every two weeks were small. However, the difference in yields of silver carp amounted to 30%. This can be attributed to the fact that silver carp feed on a lower trophic level than carp and *Sarotherodon aureus*. The higher biomass of phytoplankton rather than its production may have affected its growth rate. When half the "standard" dose of fertilizer was applied every week, keeping the total amount unchanged, no differences in yield were noted.

The foregoing discussion may be summarized from a practical viewpoint as follows:

1 Chemical fertilizers are usually required when the water temperature is over 18–20°C and the pond is under a semi-intensive management method with densities up to 3000 fish/ha.

2 Under these conditions fertilizing with a "standard" dose of 60 kg/ha single superphosphate and 60 kg/ha ammonium sulfate or liquid ammonia every 2 weeks provides good results. This amount of fertilizer can be divided into two portions applied weekly.

3 The requirement for chemical fertilizers in ponds under high-intensity management (stocking rates higher than 3000 fish/ha) and in manured ponds is not clear, especially in ponds with polyculture that includes silver carp. This warrants further study.

9.2.4 Application Methods

To ensure the uptake of mineral nutrients by phytoplankton in all parts of a pond, and to prevent nutrient loss by precipitation or release into the atmosphere due to excessive concentration in some parts of the pond, fertilizers should be distributed in the water as evenly as possible. Spreading fertilizer by casting it over the pond surface from a boat could serve this purpose, but the method is tedious. Therefore other methods are employed.

It has been found that water currents produced by even a slight wind are good distributing agents. Wind causes the upper layer of water to move in the direction of the wind, creating an undercurrent in the opposite direction near the pond bottom. These currents are used to distribute fertilizers. The fertilizers are cast on the windward side of the pond, 2–3 m from the bank, preferably when the prevailing winds are blowing. As the fertilizer dissolves it is swept away by the wind-generated current and is distributed in the pond. If this is done on the leeward side of the pond, the fertilizer is swept by the undercurrent, which will bring it into closer contact with the bottom soil and may increase nutrient fixation by absorption onto soil colloids.

A special case is the use of liquid ammonia. Liquid ammonia is often cheaper than other nitrogenous fertilizers, but it requires special precautions in its handling since it has a very strong odor and may be dangerous if inhaled or touched. Hepher (1959a) found that liquid ammonia is as efficient as any other nitrogenous fertilizer. The fertilization technique, however, is somewhat different. If the entire amount of liquid ammonia to be used is poured into a pond simultaneously, a water mass with high ammonia concentration is formed. This mass is carried slowly with the wind-generated current of the upper layer of pond water. During its movement this water mass slowly disperses, and the ammonia is dissipated throughout the water. The high concentration of ammonia is toxic to fish, but due to its strong odor, fish avoid it. The wind may, however, carry the high-concentration ammonia water mass toward a corner of the pond, trapping and killing fish. In order to avoid this, the following procedure was adopted: the amount of liquid ammonia required for fertilizing one pond is poured into an empty oil drum placed on the windward embankment of the pond. A thin plastic tubing (about 1 cm in diameter) runs from the drum, through a valve into the pond, about 2–3 m beyond the edge of the water. The outlet should not be in the area sheltered from the wind by the embankment. The diameter of the tubing should be sufficient to allow draining of the drum within 2–4 hours. Any liquid ammonia which flows into the water through the tubing is immediately caught by the current and dispersed in the pond without having enough time to form a high-concentration "mass." This way of fertilizing is not only safer for the fish but also results in better distribution of ammonia and requires less manual work.

Liquid ammonia, although it costs less than other nitrogenous fertilizers, requires capital investment for distributing equipment:

1 A stationary storage tank of about 10 m^3 (Figure 9.5).

FIGURE 9.5.   Stationary storage tank for liquid ammonia.

2   A transportation tank of about 2 m³ for distribution of the ammonia to the drums (Figure 9.6).

3   Steel oil drums, thoroughly cleaned—one for each pond.

The stationary and distribution tanks should withstand a working pressure of 0.7 atm and a test pressure of 1 atm (over ambient pressure). The inlet opening should be about 15 cm in diameter for rapid filling to avoid loss of gas. The tanks should have a safety valve 2 cm in diameter and an air inlet valve to enable air inflow during drainage. All tanks and drums should be painted white to reduce

FIGURE 9.6.   Distribution tank for liquid ammonia.

warming by sunlight. All valves should be of steel, aluminum, or PVC since copper is corroded by ammonia. Hoses for transport of ammonia from one tank to another should be made of neoprene.

When handling ammonia one should stand at the windward side to avoid the strong odor. It is best to wear goggles to protect the eyes. The vapor pressure of liquid ammonia is 0.35 atm (5 psi) at 21°C and 1.5 atm (21 psi) at 40°C. Thus, loss of ammonia will occur if the tanks remain open. When the ammonia concentration in the air reaches 16–25% it may burn on ignition. It is best, therefore, to construct stationary tanks in open, shaded areas.

9.3 MANURING

9.3.1 The Effect of Manure on Natural Food and Fish Yields

The use of manure in fishponds is not new. It has been used for millennia in China and for centuries in other countries. Woynarovich (1956, 1957) has reported on manuring of fishponds in Europe. During the 1950s it was customary to manure ponds in Israel once a year, in spring, with composted cow manure at the rate of about 5 m³/ha. This kind of manuring was not repeated during the summer for fear of anoxia. Since the technique produced no appreciable effect, it was discontinued.

After some years, manure was applied in a different way. A relatively small quantity of dry poultry waste (DPW), about 400 kg/ha, was applied every 2 weeks, or half this amount every week, during the growing season. DPW decomposes in water much faster than composted cow manure, which contains large amount of straw. The effect of DPW on the oxygen concentration was therefore smaller. The results of biweekly manuring were much more noticeable. Since fish are attracted to the DPW and are seen feeding on parts of it (possibly on feed residues), on some farms they are not fed supplementary feed on the day of manuring. This usually does not change fish yield, but it does lower feed conversion ratios.

The concept of manuring as a means of bypassing the light limitation on primary production in ponds, as discussed above, demanded a new approach. Here the organic matter is considered a link in the food chain and should therefore be treated as food. This has an important implication with regard to the kind of manure used, frequency of application, and application rate. Manure must be easily dispersable in the water, such as liquid cowshed manure or DPW, and should be applied frequently, preferably daily.

Schroeder (1974, 1975) has studied the effect of manure on production of natural food and fish yield. He found that carp, when stocked at a rate of 5000 fish/ha, grew 25–100% faster in manured ponds than in nonmanured ponds. It is difficult to evaluate the effect of manure on the natural food organisms in ponds stocked with fish since the latter graze heavily on the former. Schroeder, therefore, compared ponds with and without fish. From his results, which are presented in Table 9.2, it is clear that manure increases zooplankton and chironomid production manyfold, though this may not be apparent in ponds stocked with fish.

Similar results have been obtained by Rappaport et al. (1977). Figures for natural food organisms found in one of the analyses are presented in Table 9.3.

The increase in zooplankton and chironomids is, no doubt, mainly a result of the increase in production of bacteria and protozoa developing on the organic matter of he manure, though other ways of utilization of manure, such as the uptake of soluble organic matter, may also have affected the zooplankton.

Despite higher production of macro-food organisms, it seems they alone cannot fully account for the increased growth of fish in manured ponds. Schroeder (1978) claims that fish ingest particles of manure. Though these organic particles seem to be of low nutritional value (Kerns and Roelofs, 1977), the bacteria and protozoa that envelop the particles are of a high nutritional value. These microorganisms are "stripped off" the particles and digested by the fish. Schroeder (1978) showed that chopped straw and cotton strips digested in a water bath at 26°C increased in protein content from 1 to 15% in 7 days due to bacterial growth. Similar observations are reported by Odum and De La Cruz (1967) who examined the nature of decaying detritus and found

TABLE 9.2 Standing Crops of Zooplankton, Chironomids, and Bacteria in Manured and Nonmanured Ponds with and without Fish[a]

	Without Fish		With Fish	
	Manured	Nonmanured	Manured	Nonmanured
Zooplankton				
(g dry matter/m³)	3.3–42.4	<0.055	0.34–1.3	<0.055
Chironomids				
(100's/m²)	79–215	1–7	1–4	1–2
Bacteria				
(1000's/ml)	17–27	0.7–4.3	1.6–6.7	

[a] From Schroeder (1974).

TABLE 9.3 Natural Food Organisms Found in Water and Bottom Soil of Manured and Nonmanured Ponds at Ginossar Intensive Fish Culture Station[a][b]

Kind of Manure	Phytoplankton [2 analyses] (1000's/l)	Rotifers [3 analyses] (organisms/l)	Chironomids [3 analyses] (organisms/ 0.1 m²)
Chicken manure	16.4	1000	680
Liquid cattle manure	5.6	867	163
Corral cattle manure	3.1	247	38
Control (no manure)	2.5	170	59

[a] From Rappaport et al. (1977).
[b] Figures are averages of the analyses reported.

an increase in protein content when the particles were aged and more finely fragmented. Detritus that had recently entered the water contained 6% protein (on an ash-free basis), but by the time it had become finely divided by turbulence and decomposition, the value had risen to 24%. This is attributed by the authors to the development of a microecosystem containing bacteria, protozoa, and microalgae. Odum (1968) concluded that fine particles of detritus and the microecosystems they contain are an important factor in the feeding of mullet.

Whatever the composition of the natural food of fish, it is clear that its production in manured ponds is greatly increased and so are fish growth and yields. Moav et al. (1977) have reported on manuring experiments lasting 126 days each in 1974 and 1975. The yields obtained in their experiments in 1975 reached about 4 ton/ha on manure only (Table 9.4). Similar results were obtained in 1977 when the ponds were manured with DPW only, as compared to yields in ponds where fish were fed on protein-rich pellets (25% protein) at a total fish density of 8300 fish/ha (Wohlfarth, 1978). These results show that very high yields can be obtained by manuring.

Schroeder (1978) correlated yields obtained with good management of ponds receiving only manure with stocking density. He found a linear relationship up to 9300 fish/ha. For each fish stocked, an annual yield of 0.75 kg was obtained. Such a linear relationship also exists in fishponds using conventional pelleted feeds (Hepher and Schroeder, 1977, see Figure 7.3). For each fish stocked an annual yield of 1 kg was obtained. The yield from manured ponds was thus only 25% less than that from pellet-fed ponds. The ratio of fish gain to manure added indicated efficient utilization. For every kilogram of fish produced, about 3–3.5 kg of manure dry matter were used.

TABLE 9.4 Effect of Manure on Yield in a Polyculture System at Dor, Israel[a]

| | 1974 | | | |
| | Cow Manure Only | | Grain Diet Only | |
Fish Species	Stocking Rate (fish/ha)	Yield (kg/ha)	Stocking Rate (fish/ha)	Yield (kg/ha)
Common carp	3050	1345	3300	1778
Silver carp	1250	845	1250	1000
Tilapia aurea	1500	132	1500	162
Grass carp	330	140	330	92
Total	6130	2462	6380	3032

| | 1975 | | | | | |
| | High Stocking Rate | | | Low Stocking Rate | | |
Fish Species	Stocking Rate (fish/ha)	Manure Only (kg/ha)	Manure + Grains (kg/ha)	Stocking Rate (fish/ha)	Manure Only (kg/ha)	Manure + Grains (kg/ha)
Common carp	9050	1595	1650	4800	2194	3060
Silver carp	3080	1184	1120	1520	875	923
Tilapia aurea	5000	1072	1117	2500	686	677
Grass carp	850	270	425	420	212	263
Total	17980	4121	5212	9240	3967	4923

[a] From Moav et al. (1977).

Supplementary feeding in manured ponds noticeably affects the yield of common carp and grass carp. Silver carp, and to a certain extent tilapia, are not responsive to feeding even at high stocking rates. With an increase in density to very high levels, carp growth in manured ponds is retarded. The fish do not reach market size and the yield is lower. The effect of supplementary feed under these conditions is more pronounced.

The effect of manure in ponds heavily stocked with carp in monoculture with supplementary feeding has been studied by Rappaport et al. (1977). It proved beneficial even under high densities. The yields during 153 days at stocking densities of 20,000–45,000 fish/ha were:

	Control (Feed Only)	Liquid Cow Manure	DPW
Yield (kg/ha/day)	29.2	33.0	36.3
Extrapolated to 240 days (kg/ha)	7008	7920	8712
Feed conversion ratio	3.11	2.58	2.39

When the practice of manuring at the research stations was applied to commercial fish farms, contradictory results were obtained. While considerable increases in yields were observed on some farms, no beneficial effects were noticed on others. The following are some possible explanations for this phenomenon:

1 On some farms the maximum yield which could be obtained for a specific pond stocking rate had already been reached. Natural and supplementary food were in sufficient supply even before manure was applied, and so the fish growth rate had reached its potential. Manuring can be beneficial only if the stocking rates are increased or supplementary feed rates decreased. In the first case the yield should be higher; in the second, the yield will remain the same but the feed conversion ratio will decrease.

2 In many instances the farmers had used a very low quality of manure, containing under 5% dry matter, without increasing the manuring rates accordingly.

3 In other cases, because too much was expected of the manuring, stocking rates were increased or feeding rates were decreased excessively. When carp constituted a large proportion of the fish, this reduced total yield and increased the feed conversion ratios.

It is clear that proper use of manuring can be beneficial in increasing pond fish yields. Wohlfarth and Schroeder (1979) have reviewed the use of manure in fish farming, with a special emphasis on its relation to polyculture. They point out this beneficial effect, especially in specific economic systems such as those prevalent in Southeast Asia.

9.3.2 Kinds of Manure

The main purpose of manuring is to stimulate the growth of bacteria which develop on the organic particles and take part in the food chain. This bacterial growth depends on environmental conditions such as temperature, available oxygen, and available minerals on the one hand, and the surface area of the organic matter free for their attack, on the other. The smaller the particles of organic matter, the larger their total surface area per unit volume available for bacterial attachment and attack, and the larger the amount of bacteria produced. It is clear that any manure which produces fine, colloidal particles is more suitable for use in ponds than manure in large lumps. Since some organisms, such as bottom fauna, are limited in their movement, the manure particles should be dispersed as much as possible throughout the pond. This again means that the manure should consist of fine particles which will be distributed throughout the pond.

Any farm animal manure (or other source of organic matter such as municipal wastewater or night soil) which satisfies the two conditions—that is, fine particles, easily dispersable—is suitable for use in ponds.

Liquid cattle manure is collected from cowsheds with a slotted floor, where both manure and the urine are collected in a pit. This material contains about 10–12% dry matter. However, in many cases it is difficult to pump the manure out of the pit, and so water is added. In some cases diluted manure contains only 2–3% dry matter. Chicken manure is also readily dispersable even when added in a dry form, and it has a favorable effect on fish yields.

When manure is collected from a corral having a concrete floor and scraped into a pit, it usually loses part of the urine. Fresh manure contains 13–15% dry matter and disperses well in ponds. However, when the manure dries or is composted it loses value for manuring fish ponds. Drying and composting seem to cause the denaturation of proteins contained in the manure and the binding of the particles into larger lumps which do not disperse well in water. Poor results were obtained with these manures, even when first mixed with water and then dispersed in ponds.

Rappaport et al. (1977) found the following amounts of dry matter in suspension after diluting equal concentrations of three manures:

Chicken manure	12%
Liquid cattle manure	4.6%
Corral cattle manure	4.0%

Manure can also be used in conjunction with bio-gas production (predominantly methane). In this case, manure is first digested for gas production, and then the remaining sludge is used in fishponds. There are two main processes of digestion. The mesophillic method, with an optimal temperature of about 35°C, and the thermophillic method, with an optimal temperature of about 52°C. The sludge of both methods can be used in ponds. The thermophillic method seems to produce sludge which is higher in colloidal particles and may therefore have a better effect in ponds. Both sludges contain about 10% dry matter.

Tapiador et al. (1977) report that pig manure is used extensively in China, where 30–45 pigs are raised to supply adequate organic manure for each hectare of fishponds. In some places composting tanks are constructed so that the liquid product of fermentation flows directly into the water supply canals and to the fishponds. Woynarovich (1956,

1979) reports on the use of pig manure in Hungary. The basis of this technique is that soft, fresh manure is mixed with pond water and repeatedly spread over the entire pond surface. Good results, 500–800 kg/ha in a production period of only 150 days, were obtained in polyculture ponds stocked with common, silver, and bighead carp. Buck et al. (1979) manured fish ponds in Illinois with pig manure. The ponds were stocked with silver carp (of which about 4300 fish/ha survived), common carp (1500 fish/ha), bighead carp (409 fish/ha), grass carp (57 fish/ha), channel catfish (125 fish/ha), and largemouth bass (65 fish/ha). The ponds were manured continuously from a pigsty constructed on the embankment with a floor sloping toward the pond. One pond was manured by 39 pigs per hectare and the other by 66 pigs per hectare. Fish yields were 2971 kg/ha and 3834 kg/ha, respectively, during a period of 170 days.

The use of human waste as fertilizer is a traditional practice in China. It is estimated that "night soil" constitutes as much as one-third of the total fertilizer resources in China (Tapiador et al., 1977). However, the use of this manure in other countries is limited because of the hygienic problems involved. On the other hand, municipal wastewater is often used in ponds. A different way of manuring ponds is by raising ducks in them. The ducks continuously manure the pond and increase fish yields (see Section 9.4).

The organic matter content may be used to evaluate the various manures available. Woynarovich (1979) gives the composition of some manures used in fish culture (Table 9.5), but there may be considerable deviations from the values presented. Dry matter (or organic matter) content and the costs of manure and its transport are the main factors (besides availability) in choosing the kind of manure used. The following example illustrates this point. The price of DPW in some parts of Israel is about IL 160–180/m^3. DPW contains 50–70% dry matter. If we consider the specific gravity of this manure to be 0.7, the cost per

TABLE 9.5 Chemical Composition of Manures Used in Fish Culture (in %)[a]

Components	Pig Manure (Fresh)	Chicken Droppings	Duck Droppings	Goose Droppings
Water	71	56	57	77
Organic matter	25	26	26	14
Nitrogen	0.5	1.6	1.0	0.6
P_2O_5	0.4	1.5	1.4	0.5
K_2O	0.3	0.9	0.6	1.0
Ca	0.09	2.4	1.8	0.9

[a] From Woynarovich (1979).

ton of dry matter will be IL 325–514. Liquid cowshed manure can be obtained free. The dry matter content of this manure is about 10–12%. However, if the cost of transport of cowshed manure exceeds IL 62/ton wet weight (= IL 516/ton dry matter), it is more economical to use DPW.

9.3.3 Rates and Frequency of Application

The major limitation on the amount of manure that can be used is its effect on oxygen concentration. The decomposition of organic matter by bacteria uses large amounts of oxygen. In extreme cases this can cause fish kills by anoxia. It is important, therefore, to determine the maximum amount of organic matter that can be added without endangering the fish.

Schroeder (1974) discusses how to predict decreases in dissolved oxygen following manuring. The amount of manure that can be added depends on three main factors: the biochemical oxygen demand (BOD) of the manure, the temperature of the water, and the available oxygen supply.

The BOD of manure is specific to each kind of manure and can be determined in the laboratory. Since the decomposition of manure in summer is quite rapid and its main effect on oxygen occurs mainly during the first 24 hours, and since during daytime there is usually enough oxygen to satisfy the demand, it is the 12 hour BOD at ambient water temperature that interests us rather than the standard 5 day BOD at 20°C used in wastewater evaluation. BOD is highly correlated with the amount of dry matter present, and once this relationship is determined in the laboratory, it is easy to estimate BOD from the dry matter values.

Where winter water temperatures are low the BOD of manure is lower since its decomposition is slower. Manuring in the winter may not endanger the fish, but because of slower decomposition, organic matter may accumulate on pond bottoms to decompose at a high rate when the temperature rises in the spring, with a consequent deoxygenation and fish kill. Manuring in winter when temperatures are below 20–21°C should therefore be discouraged. During summer care should be taken that adequate dissolved oxygen (DO) is available in the ponds to sustain decomposition of the added organic matter, leaving a minimum safe DO level in the early morning when the DO daily cycle reaches its lowest point. A minimum of 2.0–2.5 ppm DO seems to be enough for carp ponds.

It is obvious that the procedure described for determining the maximum safe manure amounts cannot be carried out on the farm itself and should be worked out by the laboratory for a given set of conditions (kind of manure, prevailing water temperature, and normal DO cycle). In order to allow for differences in local conditions, a large safety margin should be maintained.

From the data given by Schroeder (1974) it seems that for liquid cowshed manure and Israeli summer water temperatures, 120–150 kg of dry matter per hectare can be applied safely at any one application. For manure containing 12% dry matter safe levels are 1–1.25 ton/ha.

The danger of deoxygenation associated with application of manure demands careful monitoring of DO by the fish farmer before and after manuring. This should be done by measuring the DO before sunrise when the oxygen concentration is at its lowest. Modern equipment makes this monitoring easier than in the past, but even when it is lacking, observation of fish behavior and birds at this hour may tell much about conditions in the pond (see Section 8.1).

The frequency of manure application affects the relationship between the amount of organic matter added and the rate of DO depletion. When manure is applied in a single dose or sporadically, the bacterial population increases unhampered to a very high level, since the development of the protozoa population, which feeds heavily on bacteria, lags somewhat behind that of the bacteria. The peak in bacterial growth corresponds with a high rate of oxygen uptake which may cause deoxygenation and fish kills. With the subsequent increase in the population of protozoa, a second peak in oxygen uptake will occur which may also be dangerous to fish.

When manure is applied periodically and frequently, constant populations of the entire food chain are established. Peak levels of bacteria after routine manuring are much lower than after infrequent application, since the protozoa and other organisms already present in large numbers graze heavily on the bacteria and limit their increase. Protozoa are also preyed upon by zooplankton and other organisms and do not reach excessive levels. In this way large amounts of organic matter can be applied and "ingested" without greatly affecting oxygen level, while at the same time producing large amounts of natural food. There is thus a constant flow of organic carbon from the manure through the food web to the fish.

Manure should be added as frequently as possible, at least daily. This is the procedure adopted in some farms in Hungary where manure is distributed five times a week at a rate (wet weight) of 30–60 ton/ha per 100 days (Woynarovich, 1979).

The amounts of manure which are added daily to ponds at the Fish and Aquaculture Research Station at Dor without causing oxygen problems reach 180–190 kg of dry matter per hectare per day. Again,

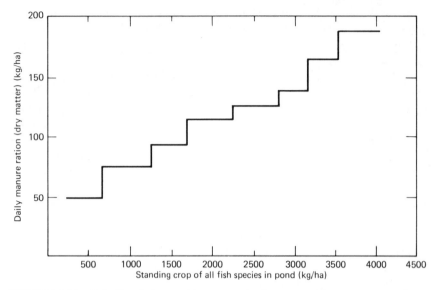

FIGURE 9.7. Relation between the amount of manure (dry matter per hectare) and the standing crop of fish. Calculated from Wohlfarth (1978).

when careful monitoring of the oxygen is not possible, a larger safety margin should be allowed for and maximum amounts of dry matter should be lower, not exceeding 100–120 kg/ha/day.

The food requirements of fish in ponds increase with increasing standing crop. There is no point, therefore, in manuring ponds with a low standing crop with a full dose. With an increasing standing crop and a resultant increase in the demand for natural food, the amounts of manure are also increased until they reach the maximum which may be safely added. Figure 9.7 gives the relationship between total standing crop and the amount of manure (dry matter) added in an experiment described by Wohlfarth (1978). The graph is calculated from his paper.

9.4 FISH-CUM-DUCK CULTURE

Three conditions for effective manuring have been mentioned above: (1) breakdown of the manure into fine colloidal particles, which seem to be highest in fresh manure; (2) frequent application; and (3) even distribution over the pond. One of the best ways to meet all three conditions is to grow the farm animals right in the pond. The last condition is best met by the culture of ducks (see Figure 9.8). Duck-cum-fish culture is not a new practice. Ducks have been kept on

FIGURE 9.8. Duck pens over a pond.

fishponds for centuries in Central Europe and China, although not on a commercial scale. It has only been since World War II that commercial fish-cum-duck culture has developed in Europe (Woynarovich, 1979).

One can regard duck culture in fishponds as an extension of the concept of polyculture. While polyculture of several species of fish having different feeding habits uses the natural food in a pond better, ducks can use food such as higher vegetation, tadpoles, and other things in a pond and along its margins. At the same time ducks contribute appreciable amounts of manure to ponds. According to Woynarovich (1979) each duck produces about 6 kg of droppings in 30–40 days or about 70 kg of manure per year.

Two methods of duck raising in fishponds have been developed: extensive and intensive. In the extensive method, the number of ducks is limited by the food they can find in the pond. Only small amounts of supplementary feed are provided to the ducks. This is the method usually employed in Europe, where 300–500 ducks are held per hectare of pond. Naturally, the amount of manure contributed to the pond and its effect on fish yield are also limited. In the intensive method the ducks are fed at the same rates as on land and held at a much higher density per unit of pond area. The amounts of manure are higher, and since some of the duck feed is lost into the pond (this is estimated at about 10%) it is also used by the fish. High yields of fish can be obtained by this method. Ling (1971) reports that in Taiwan up to 1500 ducks are raised per hectare of fishpond, and fish yield amounts to about 3500 kg/ha without additional fertilization or supplementary feeding.

In Israel fish-cum-duck culture has been attempted only on an experimental basis and for short periods, but the results have been interesting and encouraging (H. Barash, personal communication).

9.5   THE USE OF WASTEWATER IN FISH PONDS

9.5.1   The Effects of Wastewater on Fish Yield

The use of domestic wastewater in fishponds may be considered one of the best ways of manuring. Indeed, almost all experiments and commercial-scale culture where wastewater has been introduced into ponds report higher fish yields than those normally obtained in the same area. Wastewater fishponds in Munich, Germany, to which treated wastewater was introduced after a fourfold dilution with river water, yielded about 500 kg/ha, much more than obtained in regular ponds in the same region at that time (Demoll, 1926). Wolny (1962), utilizing undiluted treated wastewater effluent from the city of Kielce, Poland, in experimental ponds obtained a yield of 1318 kg/ha, which he states is a record for Poland. Much higher yields are obtained in tropical regions. In Indonesia, Vaas (1948) reports a yield of 3 tons/ha of common carp and *Puntius javanicus* in a pond receiving wastewater after diluting it first in a ratio of three parts river water to one part wastewater. Jhingran (1974) obtained a yield of 3.2 ton/ha/yr of Indian major carps in a wastewater fishpond in West Bengal, India, and recently Sreenivasan et al. (unpublished report, personal communication) concluded from experiments at Guiday, India, that 9.5 ton/ha/yr of fish (tilapia, carp, and mrigal) can be obtained in a wastewater fishpond. Hepher and Schroeder (1977) have demonstrated the benefits of wastewater aquaculture in three fishponds with a total area of 2.7 ha receiving wastewater from a village of 500 people. Table 9.6 summarizes the production in these ponds and two regular ponds in the farm with no manuring. Clearly, the added wastewater increased both the yield of fish (by 75%) and the efficiency of feed utilization (a decrease of feed conservation ratio by 53%).

In a second example, a larger proportion of wastewater was added to two larger reservoirs of 4 ha each (depth = 4 m) constructed in a

TABLE 9.6   Fish Production and Feed Conversion for Fishponds with and without Wastewater[a]

	Regular Ponds		Wastewater Ponds			Manured Ponds
Pond area (ha)	1.4	2.2	0.7	1.0	1.0	2.0
Fish yield						
(ton/ha/8 months)	4.7	4.7	8.0	8.6	8.1	7.5
Feed conversion ratio	1.8	1.6	0.6	1.0	0.8	1.1

[a] From Hepher and Schroeder (1977).

series in order to accumulate flood water during the winter. The primary treated (settled only) wastewater of a nearby town of 5000 inhabitants flowed into the first pond and spilled over into the second. The total amount of wastewater was 700–1000 m³/day. Table 9.7 gives the figures for stocking and harvesting of these ponds. Since the water was used for irrigation of cotton, the ponds were drained by early August, and the growing period was only 151 days long. It can be seen that most of the organic matter of the wastewater was utilized in the first reservoir. Higher yields could probably have been obtained if additional species, such as tilapia, had been included.

9.5.2 Methods of Application

Three methods are employed to regulate the organic load of wastewater introduced into fishponds:

Direct Application. Here the wastewater is added to the fishpond after only short primary treatment in settling tanks. In this case initial BOD is relatively high, and a larger volume of water is required to absorb and digest the organic matter in the pond (i.e., longer detention time). However, there is a lack of sufficient information, and optimal loads under different conditions have yet to be determined.

Dilution. Prior to introducing wastewater into a pond, it is diluted with 3–4 volumes of fresh water.

Pretreatment. The main purpose of pretreatment is to reduce the initial BOD, but it also removes many of the pathogens. Treatment can be done either by conventional methods (trickling filter, activated sludge, etc.) or by letting the wastewater pass through a series of lagoons. Fish are cultured in the last ponds of the series. The detention time depends on the level of pretreatment. A short detention time means increased organic load in the fishpond with consequent decreasing dissolved oxygen concentration.

When the oxygen regime in the ponds is considered, it is clear that neither initial BOD concentration in sewage effluent nor the pretreatment levels are as important as the organic load in the fishpond itself. This is not always the case when other constraints on the use of wastewater are considered (see Section 9.5.3). It seems that a reduction of pathogens and toxicants contained in water to a level which will not endanger fish or fish consumers can be reached only by pretreatment of the wastewater. Dilution may be used with weak, usually rural, wastewater. Direct application of wastewater does not seem to solve these problems and should be discouraged.

TABLE 9.7 Fish Production in Two Reservoirs[a] Supplied (in Series) with Municipal Wastewater (700–1000 m³/day)

Reservoir	Species	Stocking Date	Stocking Rate (per ha)	Avg. Wt. (g)	Harvest Date	No. of Fish (per ha)	Avg. Wt. (g)	Yield (kg/ha)	Feed (kg/ha)	Conversion Rate
1	Common carp	March 10	2750	25	July 23–	2632	748	1899		
	Silver carp	March 14	875	120	Aug. 9	616	1286	687		
						Total:		2586	2465	1.05
						Yield per day:		17.1		
2	Common carp	March 10	1500	25	July 21	1321	280	332		
	Silver carp	March 14	900	120	July 21	875	750	547		
						Total:		879	803	1.47
						Yield per day:		5.8		

[a] Each reservoir covers 4 ha and is 4 m deep (about 16,000 m³).

188

9.5.3 Constraints on the Use of Wastewater in Fishponds

Whereas the benefits of using wastewater in fishponds are obvious, a number of constraints may limit its application. In addition to the effect of organic matter on oxygen concentration and the possible development of anoxia due to an excessive organic loading, three other factors should be considered: (1) public health aspects, (2) toxicants and heavy metals, and (3) public attitudes. Each may constitute a major constraint, although suitable methods are available to overcome most objections.

Public Health Aspects

The possibility that fish may be vectors, active or passive, for the transmission of bacterial and viral pathogens is a major constraint upon wastewater aquaculture systems. Unlike warmblooded animals, fish normally do not suffer from infections of *Salmonella, Shigella,* and other enterobacteria. There seems, however, to be a controversy over whether enterobacteria can multiply in the gut, mucus, and tissues of fish and thus render the fish potential long-term vectors for human disease, or whether fish serve merely as passive carriers. N. Buras (personal communication) has found that when the concentration of pathogens in pond water is high they are found not only in the digestive tract of the fish, but also in the tissues. In this respect it is important to note that though fishponds have a high natural capacity to purify wastewater and reduce the number of pathogens, the more pathogens there are in pond water, the greater the danger of their transfer by fish.

The technique of holding fish in a densely stocked freshwater pond for a depuration period of several days prior to marketing has been the pragmatic method used as a safeguard against transmission of pathogens by fish. The actual effectiveness of and need for this process must still be clarified. It appears that the rate of depuration depends to a large extent on the degree of contamination. If pathogens have invaded the flesh of the fish, a very long period (sometimes months) will be required to depurate them.

Processing the fish at high temperatures so as to eradicate pathogens entirely may be another way of avoiding hazards to the consumer. This should be done in a plant where special measures have been taken, if necessary, to prevent hazards to workers. However, processing is quite common in sea-water fish and salmon, but not many products have been developed from freshwater fish.

Poisons, Detergents, and Heavy Metals

Wastewater is not a defined substance. It may contain substances that cause fish mortality or limit their growth, which in terms of fish farming is almost as bad. Wastewater may also contain materials such as heavy metals and pesticides in concentrations that while not lethal to the fish, may concentrate in their fiesh and create a hazard to the consumer. Some materials, such as phenols, impart an unpleasant taste and odor to fish. When planning the use of wastewater in aquaculture, special care must be taken to avoid the incorporation of materials that can harm the fish or its consumers. If possible this should be done by removing hazardous substances at their sources. A careful long-term bioassay with the fish species to be cultured can show whether a certain effluent is harmful to fish culture.

One of the most troublesome substances in municipal wastewater is detergent which originates from dispersed, multiple sources in domestic wastewater. Detergents may be toxic to fish, affecting the mucus coating and rendering the fish more susceptible to attack from diseases and parasites. High concentrations of detergents—10–18 ppm alkylbenzene sulfonate (ABS) have been reported by Hepher and Schroeder (1977) for the wastewater of Haifa. This is above the lethal concentration for common carp, which is about 10 ppm ABS (Hepher, unpublished report). Sublethal concentrations of the hard detergents may still affect fish and retard their growth.

The accumulation of heavy metals and pesticides in fish cultured in wastewater should also be considered and studied. The growth rate of fish in ponds is quite rapid, and the duration of the growth period is usually short, thus lessening the chance of accumulating heavy metals, unless the concentration of these metals is very high. Kerfoot and Redmann (1974) have studied the accumulation of heavy metals in oysters cultured in a marine aquaculture system receiving secondary sewage. They found that the inherent elevated metal concentration in domestic wastewater did not appear to pose a threat to shellfish and that there was no accumulation of these metals in the shellfish. A. Perry (personal communication) has found that fish cultured in wastewater for 70 days do show a certain accumulation of heavy metals and organic residues, but the contents of these compounds in the fish were within the accepted standards for safe use.

Public Attitudes

No less serious than sanitary problems is the attitude of the public toward using wastewater in fishponds. There is no point in culturing fish in wastewater ponds if the fish cannot be sold. This attitude is not necessarily related to the hygienic safety associated with consuming these fish. Even when the fish are safe for use there will be public opposition, although this can be somewhat reduced by education and dissemination of correct information. The best way to overcome public opposition seems to be to remove direct contact between fish and sewage. This can be done by having some intermediate stage between the raw sewage and the fish culture pond, such as pretreatment plants or an oxidation lagoon. The effluent then becomes "treated" or "reclaimed" water rather than "sewage." Processing of fish also contributes to the dissociation of the product from the wastewater fishpond.

10 Nutrition and Feeding

10.1 PRINCIPLES OF SUPPLEMENTARY FEEDING

In order to grow at their potential rate fish require food that will sustain them and provide for their growth. In short, growing fish need a complete diet. In raceways and cages where little, if any, natural food is available, a complete diet must be provided by the fish farmer. These are usually diets rich in protein and vitamins, and they are therefore quite expensive. The costs of feed in intensive culture systems may comprise 50% or more of total costs (Collins and Delmendo, 1979). This limits the use of complete diets to luxury fish such as trout. Warmwater fishes, which are usually less valuable, are produced in systems, mainly ponds, where natural food is available. The cost of natural food production is low (if fertilizers and manure are used). However, the yield which can be obtained from natural food alone is also low, and sometimes it is insufficient to cover the fixed costs related to pond area (Tal and Hepher, 1967). The profitability of fish culture systems often depends on supplementary diets, the compositions of which are simpler and less expensive than complete diets. In order to be economically beneficial, the efficiency of supplementary diets must be as high as possible. This will depend to a large extent on the level of feeding and the composition of the diet.

Feeding level and composition of the complete diets are determined by the nutritional requirements of the fish. Amounts and composition of the supplementary diets, however, are determined both by the total food requirement of the fish population and by the amount and composition of the natural food that the fish are getting from the pond. It is the natural food that is supplemented. Both these factors change constantly. This makes the determination of quantity and quality of the supplementary diet very difficult.

Direct evaluation of the natural food levels available is almost impossible. Each fish species has its own natural food, and this may change with fish age and especially during times of food scarcity. Thus, the determination of the phytoplankton, zooplankton, and benthos biomass or that of specific organisms in the pond will not necessarily represent the food available to the fish. Moreover, even if the specific food of the fish is known, the biomasses of those food items does not represent the food available. Production rates are more relevant here, but no reliable method for the estimation of secondary production has as yet been developed.

The error in estimating biomass of natural food as an index for production can be seen from Table 9.2, which gives biomass of some components of natural food found in manured and nonmanured ponds (Schroeder, 1974, 1975). Gain in carp yield in manured ponds was about double that of nonmanured ponds. This was no doubt due to increased production of natural food in the manured ponds, but these ponds stocked with fish did not reflect increased production in the biomass of natural food because most of the organisms produced in the manured ponds were consumed by the fish. It was only in ponds without fish that the effect of manure on natural food could be seen. Lellak (1957) secluded protected areas within a pond from fish and measured natural food productivity. However, this has been criticized by Hruška (1961) on the grounds that: (1) there is a difference in the production rate between the protected and unprotected area; and (2) there is a partial migration of larvae from (and into) protected areas. It seems, therefore, that the only method for estimating natural food consumption by fish is an indirect one. It may be done by measuring the standing crop of the fish which this natural food can sustain and the fish growth rate. Naturally, this cannot give the answer for an individual situation, but only the average for a pond, a productivity level, or a geographical area.

The main difficulty in determining feeding level and composition of supplementary diets lies in the fact that the standing crop of fish, and thus its food requirement, varies constantly. With increases in both the density of fish per unit area and in the weight of the fish, the food requirement of the population increases. This increase is usually followed by an increase in natural food. However, the increase in production of natural food is slower than the increase in food requirement. In many cases increase in density is associated with a decrease in the production of the natural food due to overgrazing or to changes in environmental conditions (Merla, 1966). This means that the ratio of consumed natural food to the overall food requirement decreases with increased fish standing crop. When the density of fish per unit area increases, each fish may get a smaller average portion of the natural food. When, however, fish density is constant but the weight of fish increases, each fish may get the same amount of natural food but this covers a smaller portion of its requirement. In both cases a deficit develops which must be covered by supplementary feed.

Natural food does not necessarily supply the various food components such as carbohydrates, proteins, or vitamins in the same proportions as they are required by fish. With the decrease in the ratio of natural food to total food requirement, the deficit in these components does not have to be the same. The case of the common carp can serve as a good example.

Some studies have shown that the nutritional requirements of common carp can be satisfied by a diet containing 36–40% protein and rich in energy (Steffens, 1966; Kaneko, 1969; Sin, 1973a, 1973b; Shiloh and Viola, 1973). Proteins comprise about 50–60% of the calorific value of natural food. This means that when carp are provided only natural food, they use part of the protein for energy. When natural food becomes scarce, it is energy and not protein which is first lacking. Supplementary feed should cover this energy deficit first and need contain mainly sources of energy. Cereal grains rich in carbohydrates are usually used as feed at this stage. It may seem illogical that by feeding carp in ponds with carbohydrates one gets fish protein in return. The truth is, of course, that the growth is supported by the natural good protein "released" by the addition of the energy.

With increasing standing crop of carp, a point is reached where natural proteins are no longer sufficient to sustain maximum growth. Supplementary protein should then be added to the diet. The higher the standing crop of fish, the larger the deficit in natural protein, and the higher should be the protein content of the diet. Hepher et al. (1971) and Hepher (1975) have shown that there is no difference in carp growth when the carp are fed cereal grains (sorghum) or a protein-rich pellet containing 22.5% protein up to a standing crop of 800 kg/ha. There is no difference between the effect of pellets containing 22.5% protein and pellets containing 27.5% protein up to a standing crop of 1400 kg carp/ha. Since with the addition of proteinous ingredients, such as fish meal, less carbohydrate-rich ingredients can be included in the diet, high levels of protein in the diet may reduce its energy/protein ratio and at some point oil must be added to retain this ratio (apart from the effect of oil as a source of essential fatty acids). The same arguments are true for the inclusion of vitamins and minerals in the diet. No effect of dietary vitamins has been noticed in feeding experiments at the Fish and Aquaculture Research Station at Dor below a standing crop of 2.4 tons of carp per hectare, but vitamins have been found to affect growth over this standing crop.

It is obvious that there are "critical standing crops" beyond which natural food does not adequately supply all the nutritional components required for growth. When a deficit develops, supplementary feed has to supply the lacking components in increasing amounts along with the increase in standing crop. If studied further, it will probably be found that these critical standing crops differ not only with regard to the major food components such as energy and protein, but also with regard to individual amino acids or specific vitamins.

Since critical standing crops depend on the ratio between the supply of natural food and the nutritional requirements of the fish, it is obvious that the critical standing crop will be different in ponds with different natural productivities and that it can change as a result of fertilizing or manuring. A certain standing crop in a nonmanured pond may require a supplementary diet rich in protein, while the same standing crop in a manured pond may produce the same fish yield on cereal grains only.

From the above discussion it is obvious that one cannot speak of a certain fixed composition of the supplementary diet. It may differ not only between different fish species, as one may expect, but also between different standing crops of fish of the same species, or between the same standing crop, at different productivity levels.

A number of conclusions can be drawn from the above points:

1 To determine the composition of the most efficient supplementary diet, one must ascertain the relevant critical standing crop for each of the species dealt with. Since this cannot be done for a single pond or farm, standards should be set by experiments and analyses of average growth rates in the farms, which will apply for an entire geographical area and for various productivity levels.

2 Since the composition of the diet is changing continuously with increasing standing crop, it would require an infinite number of diets if the maximum utilization is sought. This is, of course, impossible, but a partial solution can be found in one of the following possibilities:

a. A limited number of diets of increasing nutritional value are prepared. Each of these will cover the nutritional requirements up to a given critical point, reached at a known standing crop, when it is switched to the next one.

b. A complete diet, or a diet rich in protein, is "diluted" with a diet rich in carbohydrates. This is the usual practice in Israeli fish farms (see Section 10.3). Up to about 800 kg/ha of carp, the fish are fed cereal grains only. Above that, a mixture of cereal grains and pellets containing 25% protein are given at changing proportions ($\frac{3}{4}:\frac{1}{4}$, $\frac{1}{2}:\frac{1}{2}$, $\frac{1}{4}:\frac{3}{4}$) with increasing standing crop, At standing crops of over about 1800 kg carp/ha only pellets should be used.

10.2 FEEDSTUFFS

10.2.1 Simple Feedstuffs

Most of the simple feedstuffs used for fish are rich in carbohydrates. Some of these are relatively inexpensive farm byproducts such as rice bran. Protein-rich feedstuffs such as oil cakes are only rarely used, mostly in Eastern Europe. Plants are used extensively in the Far East, especially China, for feeding grass carp. While compounded feeds are readily accepted by most pond fish species, the simple feedstuffs are often not accepted by some pond fishes. Whole cereal grains such as sorghum are readily eaten (after becoming softened by the water) by common carp, but not by tilapia. Grass is consumed by grass carp, but not by common carp or most other pond fishes. Some fish, such as silver and bighead carp, take only very fine feed.

Carbohydrate-Rich Feedstuffs

Carbohydrate-rich feeds are fed mainly to common carp. Any cereal grain or legume seed can serve as feed. The latter are richer in protein. The most common feed in Israel is sorghum (milo), but other grains such as wheat (shrunken or broken wheat grains rejected by flour mills), bitter lupine, and others have been used. Rye and barley are commonly used for feeding common carp in Europe, and rice bran is used in the Far East. No doubt other carbohydrate-rich feedstuffs can also be used. During World War II, raw and cooked potatoes were used for feeding carp in Germany, although feed conversion ratios were high (Wunder, 1937; Demoll, 1940a, 1940b).

One should bear in mind that feeding high levels of carbohydrate to carp may cause a considerable accumulation of body fat. During 1968 Israeli fish farmers had difficulty selling fish. Consequently, the fish were kept at high densities in storage ponds in which they were fed sorghum so as not to lose weight. They soon accumulated high levels of body fat (up to 34% of their wet weight). This reduced their market value considerably. It was found that sorghum produces more body fat than wheat does, probably due to the higher protein content of the latter.

Protein-Rich Feedstuffs

Protein-rich feedstuffs can be classified into two groups according to their origin: (1) from vegetable sources, such as oil meals; and (2) from animal sources, such as trash fish and meat and poultry processing byproducts. Fish meal is not fed directly, but rather as an ingredient of compounded feed.

Oil meals usually contain 35–45% protein. When fed alone they do not produce good growth in carp. Müller (1959) found that when sunflower oil meal was fed directly, carp grew less than they did on wheat. A better result was obtained when the oil meal was added as fertilizer to ponds where fish were fed on wheat, rather than when the meal was used directly as feed. Cottonseed oil meal was especially poor. This may have been due to the high protein/energy ratio and/or the low biological value of the protein. Cottonseed meal contains gossypol which may also inhibit growth.

Poultry processing byproducts contain the entrails, heads, feet, and shanks discarded when the poultry is dressed for marketing. These parts must be chopped into smaller particles by a mechanical chopper and then fed to fish in special floating wooden frames. The protein in these byproducts, being of animal source, has a high biological value, but here too the protein/energy ratio is high. For optimal utilization it should be given with an energy source such as cereal grains. Since in most cases the available amounts of these wastes are small, the usual practice on farms which use them is to feed at a rate of about 20 kg/ha, making up for the rest of the supplementary feed with cereals and, at high standing crops, with pellets.

Grasses and Water Plants

Some fish, such as grass carp (*Ctenopharyngodon idella*) and *Tilapia rendalli,* can feed on higher vegetation. One of these, the grass carp, is the major pond fish in some regions of mainland China where it comprises about 55% of pond fish stock. The main fish feed used in these regions is plant material produced on the farm; supplementary feeding with high-protein feedstuffs is not practiced. The Chinese say, "Feed one grass carp well and you feed three other fish" (Tapiador et al., 1977). The plants used as feed in China and Southeast Asia are usually elephant grass, oil seed plants, and vegetables that are grown on pond embankments. The feed conversion ratios are quite high (60–70), but since the price of the plants is very low, their use is economical. It is obvious that in large ponds the embankment area in relation to water surface is small and may not supply the plants necessary to feed grass carp. In Israel grass carp have been cultured as secondary fish in polyculture systems. Their density does not usually exceed 500 fish/ha. It has been observed, however, that their growth rate on regular supplementary feedstuffs (cereal grains, protein-rich pellets) is not as high as reported in the literature for ponds in which they are fed grass.

A recent study by Porath et al. (1979) has found that the growth of grass carp can be stimulated by feeding duckweed (*Lemna* sp.). Grass carp prefer the soft duckweed over high fiber plants. Since this plant reproduces mainly by budding, it is capable of producing a high yield per unit water area. It was found that a well-manured 1 ha pond can produce 10 tons of duckweed per week (D. Porath, personal communication). When fed to grass carp, duckweed produced a much higher growth rate than protein-rich pellets and the protein was utilized much better. An experiment where grass carp were stocked at the rate of 2500 fish/ha in ponds and cultured in three different treatments showed the following results:

Treatment	Gain (g/fish/day)
No feed	1.5
Feeding with high-protein pellets (25% crude protein, 15% fish meal)	2.4
Duckweed	4.3

The feed conversion ratio of duckweed—if calculated on a wet weight basis—is quite high, about 37 kg of feed for each kilogram of fish gain. However, the dry matter content of the duckweed is low, and when the conversion ratio is calculated on the basis of dry matter or by protein content, a much better utilization of duckweed (i.e. a lower feed conversion ratio and a higher PER.*) than of protein-rich pellets is found. The experiment showed these results:

	Dry Matter Feed Conversion Ratio	PER*
High-protein pellets	3.2	1.1
Duckweed	1.9	2.6

Since duckweed has worldwide distribution and has different ecotypes adapted to specific environments, it has yet to be determined which ecotype will produce the highest yields in ponds. When duckweed is naturally available in the vicinity of the pond it is worthwhile to use it as feed for grass carp instead of, or together with, pelleted feed.

* Protein efficiency ratio = gain in body weight per amount of protein consumed.

10.2.2 Compounded Feed

The only way to feed fish diets composed of a mixture of different ingredients which will have the proper ratio between energy and protein and which will include, if necessary, vitamins and minerals, is to compound these ingredients. Two common forms of compounded feed are pellets and dough. Some studies comparing the efficiency of these two application methods have been conducted by Erokhina (1959). She found that much of the feed in dough becomes inaccessible to fish either by simply disintegrating and dispersing in the water or by entering into solution. Pellets were very effective as compounded fish feed. In her experiments, fish fed pellets reached 600–650 g at the end of the experiment, while those fed the same amount of feed in the form of a dough reached only 520–540 g. Feed conversion was 2.2 on pellets compared with 3.0 on dough. Recently, however, Fijan et al. (1976) have suggested using a mixed dry diet in a bag that has an opening cut out at its bottom and is floated on the water. When the mixture gets wet it becomes doughlike and does not fall out of the bag. The fish nibble the dough, and the bag becomes a kind of feeder. This method has yet to be used under farm conditions, and the loss of nutrients, if any, must be ascertained. In practice the pellet form is more widely used for fish feed.

Two aspects of pelleted feeds are important: (1) their nutritional value and (2) the technology of their production.

From a nutritional viewpoint pelleted feeds can be complete diets, in which case they must contain all the components that maintain the fish and support maximum growth. These diets are often rich in proteins. Cold-water fish diets (mainly trout diets) contain 40–45% protein. Warm-water fish diets can also contain a high percentage of protein. The diet used for carp at the Tanaka Farm in Japan contains 42% protein (Tanaks, personal communication). Sin (1973a) has found that the best results with carp fingerlings are obtained with 38% protein. As explained above, supplementary feed may contain much less protein, since much of the required protein is supplied by the natural food. Catfish feed contains about 32–35% protein, and sometimes even less (25%). In Israel a pelleted fish diet usually contains 25% protein. Some diets, which are used with relatively low standing crops of fish, contain even less protein (18%).

Many experiments with catfish and carp have shown that fish meal has a special importance in the diet and that whenever the fish meal is replaced by a vegetable protein source, such as soybean oil meal, the growth rate of the fish decreases. Standard carp pellets in Israel contain 15% fish meal. At low standing crops when the demand for protein is relatively low, an increase in protein with ingredients other than fish meal can compensate for low fish meal. This procedure must be critically examined, however, from an economic point of view. At a high standing crop when higher protein levels are required, the content of fish meal should not be lower than 15%.

Energy content and source are also important. Energy is required both for maintenance and for anabolic growth processes. There is some difference between coldwater fish such as trout, which cannot use carbohydrates well for this purpose and depend mainly on proteins and fats, and warm-water fish such as carp, which can use them well. Carbohydrates usually comprise an appreciable part of a warm-water fish diet. It should be noted that too little energy from carbohydrates or fats in the diet causes the utilization of the protein for energy rather than for growth. Though proteins are good sources of energy for fish, they cost much more than carbohydrates and should not be wasted. Many studies have shown that the addition of fat to the diet has a "sparing effect" on protein in trout diets (e.g., Reinitz et al., 1978) as well as carp diets (Sin, 1973b). This means that the protein in these diets is better utilized for growth. On the other hand, too much energy may lead, especially in carp, to the accumulation of body fat which reduces market quality.

Some experiments and observations have shown that carp feed should contain about 3 kcal of metabolizable energy per gram. When lower levels of dietary protein are used at relatively low standing crops of carp, this energy is supplied by cereal grains rich in carbohydrates. However, when the level of protein in the diet increases, the amount of carbohydrate decreases accordingly. It is impossible to add more energy by cereal meals because of their bulkiness. Fat, which contains two and one-half times more energy than the carbohydrate, is therefore used. Fat also has an important role in supplying the necessary essential fatty acids—the linoleic and linolenic acids. As for the form of added oil, Viola and Rappaport (1977) have found that for carp acidulated soapstocks are as effective as any fat and that there is no difference between different fats except for a slight difference in the composition of the fish's body fat which resembles, to a certain extent, the dietary fat (Viola and Amidan, 1978).

Vitamins are required in the complete diets of all fish. However, most carp supplementary diets do not contain added vitamin mix since the vitamins are supplied in sufficient amounts by natural food. Some experiments (Hepher, unpublished) have shown that supplementary vitamins do not significantly improve growth unless the fish standing crop exceeds 2.4 ton/ha., at which point a deficit in certain vitamins develops and their addition affects the growth rate. Minerals are also not added to carp feed though recent experiments by Nose and Arai (1979) show that dietary phosphate is indispensable and that the addition of phosphate to diets of carp has affected growth both in aquaria and ponds at high standing crops.

From a technological viewpoint, the main requirement for pellets is stability both when dry and when in the water. The stability of dry pellets is important since the dust created by their disintegration is lost and cannot be used by most fish. No less important, however, is pellet stability in the water. The pellet must be stable until it is consumed. This is especially important with carp which are "lazy" feeders and usually take 20–30 minutes to eat their food. The pellets should, therefore, remain stable for that period. The recent greater use of demand and automatic feeders means that water stability is less important since it seems that the fish take up the pellets as they fall from the feeder.

High water stability of pellets can be achieved in two ways: (1) by increasing the temperature during production of the pellets and (2) by using binders. High temperature, especially when the mixture is moist, gelatinizes part of the starch and turns it into a binder and a partly waterproof coating. This can be done by precooking the mixture and extruding it under pressure. The resulting pellets usually expand and therefore float. They are also very stable in water. Many catfish farmers prefer these pellets since they can see the fish coming up to feed and can estimate the size of the fish without weighing them. These pellets are expensive, however, and it is doubtful whether this extra cost is justified.

Processing temperature may also be raised by increasing the friction of the mixture while the pellets are produced in the die. Coarse particles in the mixture increase the heat of this friction. Due to the gelatinizing process, the outer surface of the pellets becomes glossy. Incorporating very fine particles in the mixture seals the small hollows in the glossy surface, through which water can be absorbed into the pellet, thus increasing stability. Further increases in temperature can be achieved by increasing the thickness of the die. This of course increases wear and tear of the costly die and decreases the production of the pellets, both of which result in increased manufacturing costs. Feed mills are usually reluctant to use this technique.

When long-lasting stability is required, such as with feed for crustaceans, agar or alginates are usually used as binders, but this sometimes requires a complicated method of preparation which is not always applicable for commercial manufacturing of the pellets. Wheat gluten is a much more widely used binder. Processed wheat gluten, when incorporated into fish pellets at about 5%, gives the pellets a water stability of 20–30 minutes. The same effect is achieved, however, when gound wheat meal is one of the pellet ingredients, thus providing both energy and binding. For this purpose a low-grade wheat, broken or shrunken and rejected by the flour mills, can be used. To achieve good binding pellets should contain at least 60% wheat.

Both dietary and technological considerations impose certain constraints on the composition of pellet-compounded fish diets. The most common pelleted diets in Israel today are:

Cereal Mixture.   This is a pelleted mixture of wheat and other cereals which contains about 10–11% protein. It is intended to replace cereal grains when mechanical feeders are used.

A Diet Containing 18% Protein.   This usually contains:

   80–90% wheat (finely ground)
   5–10% fish meal
   5–10% soybean meal

A Diet Containing 25% Protein.   The composition of this diet is approximately as follows:

   60–70% wheat, sometimes partly replaced by other cereals when mechanical feeders are used.
   15% fish meal, containing about 65% protein. When herring meal, containing about 72% protein, is used, the fish meal percentage is somewhat reduced.
   15–25% soybean meal.
   3–4% soapstock oil

10.3 FEEDING LEVELS

Food requirements for maintenance and growth increase with increasing fish size (weight), but the relative food requirements, that is, the requirement per unit weight, decrease with increasing fish size. If calculated accordingly, a 100 g fish will require less food for maintenance and sustaining maximum growth than a fish of 1000 g. However, 10 fish of 100 g (total 1000 g) together will require more food than the single fish which equals their total weight. This means that if the feeding rate is calculated per unit of fish weight (in this case 1000 g), it will decrease with increasing individual fish weight.

A common method of calculating daily feeding level is on the basis of a percentage of the live weight of the fish. Since this expresses the amount of food relative to a unit weight of fish (grams per 100 g of fish), it follows from the above that when all the food requirements are supplied by the given diet (as in the case of trout) this percentage will decrease with the increase in weight of individual fish. This is quite pronounced in feeding charts for trout or salmon. The situation is different in stagnant water ponds, where an appreciable part of the total nutrients is supplied by natural food produced in the pond. The calculation of feeding level for supplemental feed therefore becomes a complicated task. This is why feeding charts for fish in ponds are usually based less on assumptions and calculations and more on empirical experience. The most common chart used in Israel has been developed by Marek (1975). This chart (Table 10.1) takes mainly the carp into account since they are the main consumers of supplemental food in polyculture systems. Tilapia use natural food better; therefore only half on the standing crop of tilapia is taken into account for calculating feeding levels. The other fish species do not use supplemental feed except for that left over by carp and tilapia, or except indirectly, as organic detritus. The chart takes into account the change from sorghum to protein-rich pellets with the increase in standing crop as explained in Section 10.1. Details of the changeover are as follows:

Up to 700 kg/ha	Sorghum only
700–1200 kg/ha	75% sorghum, 25% pellets (25% protein)
1200–1500 kg/ha	50% sorghum, 50% pellets (25% protein)
1500–1800 kg/ha	25% sorghum, 75% pellets (25% protein
Over 1800 kg/ha	Pellets only (25% protein)

When pellets containing 18% protein are used this can be considered to be equal to the mixture of 50% sorghum and 50% pellets of 25% protein.

Marek (1975) considers that at high densities of fish natural food plays a lesser role, so an increase in the amount of supplemental feed is required. This also causes a larger difference in the feeding level, when calculated as a percentage of live weight, between small and large fish. At high densities Marek prefers to calculate feeding level per individual fish and multiply the amount by fish density. The chart for higher

TABLE 10.1 Marek's Feeding Chart for Intermediate Densities (in kg/ha)[a]

Standing Crop[b] (kg/ha)	18–20°C				20–25°C				>25°C			
	Density[c]				Density				Density			
	(<1800/ha)		(>2200/ha)		(<1800/ha)		(>2200/ha)		(<1800/ha)		(>2200/ha)	
	S[d]	P[d]	S	P	S	P	S	P	S	P	S	P
200	6	—	4	—	10	—	6	—	16	—	10	—
400	10	—	6	—	16	—	10	—	22	—	16	—
600	14	—	8	—	22	—	14	—	28	—	22	—
800	13.5	4.5	7.5	2.5	19.5	6.5	13.5	4.5	25.5	8.5	19.5	6.5
1000	15	5	9	3	22.5	7.5	15	5	30	10	22.5	7.5
1200	16.5	5.5	10.5	3.5	25.5	8.5	16.5	5.5	34.5	11.5	25.5	8.5
1400	12	12	8	8	19	19	12	12	26	26	19	19
1600	6.5	12.5	4.5	13.5	10	30	6.5	19.5	15.5	43.5	10	30
1800	7	21	5	15	10.5	31.5	7	21	5.5	46.5	10.5	31.5
2000	—	30	—	22	—	44	—	30	—	66	—	44
2200	—	32	—	24	—	46	—	32	—	68	—	46
2400	—	34	—	26	—	48	—	34	—	70	—	48
2600	—	36	—	28	—	50	—	36	—	72	—	50
2800	—	38	—	30	—	52	—	38	—	74	—	52
3000	—	40	—	32	—	54	—	40	—	76	—	54

[a] From Marek (1975).

[b] Below a standing crop of 700 kg/ha (fish fed on cereals) this includes the biomass of carp only. Above this standing crop, half of the biomass of tilapia is added.

[c] Below a standing crop of 700 kg/ha (fish fed on cereals) the "density" includes the number of carp only. Above this standing crop (fish fed on cereals and protein-rich pellets) the "density" consists of the number of carp and tilapia but not of other species. This also applies to the 20–25°C and >25°C ranges.

[d] S = sorghum; P = pellets containing 25% protein (15% fish meal).

TABLE 10.2 Marek's Feeding Chart Calculated by the Individual Weight of the Fish (g/fish/day)[a]

Carp

Fish Weight (g)	2000–4000				4000–6000				6000–8000				8000–12,000			
	S[b]	P[b]	T[b]	%[b]	S	P	T	%	S	P	T	%	S	P	T	%
20–50	1	—	1	2.9	2	—	2	5.8	1	1	2	5.8	2	1	3	8.7
50–100	2	—	2	2.7	3	—	3	4.0	3	1	4	5.4	2	2	4	5.4
100–200	5	1	6	4.0	6	2	8	5.3	5	4	9	6.0	2	7	9	6.0
200–300	7	3	10	4.0	5	6	11	4.4	2	10	12	4.8	2	8	10	4.0
300–400	5	6	11	3.1	3	10	13	3.7	3	11	14	4.0	4	11	15	4.3
400–500	5	9	14	3.1	3	12	15	3.3	3	13	16	3.6	3	14	17	3.8
500–600	4	11	15	2.7	3	13	16	2.9	3	14	17	3.1	3	15	18	3.3
600–700	3	12	15	2.3	3	14	17	2.6	3	15	18	2.8	4	15	19	2.9
700–800	4	12	16	2.1	4	14	18	2.4	4	15	19	2.5	4	16	20	2.7
800–900	4	13	17	2.0	3	15	18	2.1	3	16	19	2.2	3	17	20	2.4
900–1000	3	14	17	1.8	4	15	19	2.0	4	16	20	2.1	4	17	21	2.2
1000–1100	4	14	18	1.7	4	16	20	1.9	4	17	21	2.0	4	18	22	2.1
1100–1200	4	14	18	1.6	4	16	20	1.7	4	17	21	1.8	4	18	22	1.9

Tilapia

Fish Weight (g)	Polyculture with Carp		Monoculture	
	P	%	P	%
5–10	0.4	5.3	0.5	6.7
10–20	0.6	4.0	0.8	5.3
20–50	1.3	3.7	1.6	4.6
50–70	1.6	2.7	2.0	3.3
70–100	1.9	2.2	2.4	2.8

Fish Weight (g)	Polyculture with Carp		Monoculture	
	P	%	P	%
100–150	2.2	1.8	2.7	2.2
150–200	2.5	1.4	3.0	1.7
200–300	3.0	1.2	3.7	1.5
300–400	3.6	1.0	4.5	1.3
400–500	4.2	0.9	5.2	1.2
500–600	4.8	0.9	6.0	1.1

[a] From Marek (1975).

[b] S = sorghum; P = pellets; T = total; % = percentage of live body weight.

densities is given as Table 10.2. As in the previous case, when tilapia are in polyculture they get half the amount of carp; but tilapia in monoculture receive more. Here again there is an increase in protein-rich pellets (25% protein) concurrent with a decrease in sorghum. However, even in the high standing crops Marek still gives some sorghum to supply energy. Lately fat has been added to protein-rich pellets, and it can replace sorghum in these high standing crops.

In both feeding charts there is an appreciable reduction in feeding level calculated as a percentage of live weight of the fish. In the higher density, there is a reduction from 8.7% for a fish of 20–30 g to 1.9% for a fish of 1100–1200 g. Even when density is intermediate the feeding level changes (at over 25°C) from 5–8% to 1.8–2.5% when the standing crop increases. Two factors may have been underestimated in the calculation of Marek's charts: (1) Carp tend to accumulate fat as they get larger. From experiments, it seems that this fat accumulation is directly correlated to fish weight. This means that higher amounts of feed are required by larger fish to provide for fat accumulation. (2) Natural food seems to play a more important role than assumed in the charts. At lower standing stocks of fish the total food requirement is still low, and natural food plays an appreciable part in supplying this requirement. The requirement for supplemental feed may therefore be low, that is, a smaller percentage of body weight than given in Table 10.2. No less important is the situation when the standing crop is increasing. For a given stocking density the absolute total amount of food—comprised both of natural and supplemental feed—must increase with increasing standing crop (even though the relative amount of food, that is, the amount as a percentage of the live weight of the fish, decreases with increasing standing crop). It may be assumed that the amount of natural food produced in the pond increases with the increase in standing crop and with the increase in rate of supplementary feeding associated with it, due to the manuring effect of the undigested portion of feed. However, it is doubtful whether natural food increases at the same rate as the increase in food requirements. The relative proportion of the supplementary feed must increase with the increase in the weight of the fish (or the standing crop) to make up for the decreasing proportion of natural food. This means that the amount of supplementary feed supplied must be higher than the increase in the total nutritional requirement of the fish. Therefore even when calculated as a percentage of body weight the supplementary feed does not seem to decrease much with fish size. A feeding level of 4–5%, which is quite common on many commercial farms, seems to be not much off the mark, although more experiments are required to determine the optimal feeding level at different fish weights.

11 ——————— Hazards and Diseases

11.1 ANOXIA

Oxygen deficiency and the resultant killing of fish due to suffocation (anoxia) are the primary cause of losses in fishponds. Since anoxia occurs more frequently in productive ponds which are rich in organic matter, it is more common on intensive fish farms, and the losses associated with it can be high. Great efforts should be made, therefore, to prevent anoxia. In order to do so, the factors affecting oxygen concentration in water and its daily cycle must first be understood.

11.1.1 The Oxygen Cycle in Water

Except for a number of organisms that either obligatorily or facultatively breath air, such as some insects, amphibians, snakes, and air-breathing fish, all other aerobic organisms in ponds get the oxygen they need for respiration from its solution in water. Their efficiency of utilization of atmospheric oxygen for respiration is so low as to make oxygen, in fact, unavailable from this source. Though our major concern is the fish population in the pond which can, if stocked at a high rate, become the major consumer of oxygen, other groups of organisms also share the available dissolved oxygen (DO). Among these are communities of plankton, benthos, and bacteria. Even phytoplankton, which produces oxygen by photosynthesis, respires at the same time, and in the dark it may become one of the major oxygen consumers.

The rate of oxygen uptake depends primarily on the abundance of organisms in the water and on the water temperature which affects their rate of metabolism. Very often an increase in the amount of organic detritus in the water will cause a rapid increase in the heterotrophic bacterial populations that utilize and decompose the organic matter, and subsequently an increase in the populations of organisms, such as protozoa and larger zooplankton, that directly or indirectly feed on bacteria. This increases DO consumption, especially in warm water.

There are two sources of oxygen in ponds. The most important is photosynthesis, mainly by phytoplankton algae. The other source is the atmosphere, which contains about 21% oxygen.

The amount of oxygen that can be dissolved in water from these two sources is limited and depends mainly on (1) the pressure of oxygen on the water surface and (2) the temperature of the water. Since the partial pressure of oxygen in the atmosphere is quite constant and equals about 0.21 atm, under natural conditions saturation level will be determined mainly by temperature. The higher the temperature, the lower the DO saturation point. Table 11.1 gives the DO concentration at saturation in distilled water at various temperatures. These levels are referred to as 100% saturation, and the actual oxygen concentration is sometimes expressed as a percentage of saturation. Remember, however, that these levels are related to a certain temperature. Saturation in summer at 32°C, a temperature quite often attained in shallow ponds in warm climates, is 7.32 mg/l, which is only 64% saturation at 8°C, which can be reached in winter. This should be of special importance when critical DO levels are considered. For example, 7% DO saturation at 8°C means about 0.8 mg/l, which though very low and below the level of convenience, can sustain carp for several hours. But the same 7% DO saturation at 32°C is 0.5 mg/l, and the death of carp will occur within a short period.

When DO is below saturation, oxygen will dissolve from the atmosphere. The greater the difference, the quicker the solution. When the DO is above saturation level, oxygen will be lost to the air. The greater the difference, the quicker the loss. However, oxygen transfer from or to pond water is done through the interface between the atmosphere and water, and this layer becomes a barrier for transfer. Once a thin water film at the interface becomes saturated with oxygen, further transfer will stop until this oxygen is dissipated in the water body and

TABLE 11.1 Solubility of Oxygen in Distilled Water from Moist Air at 1 atm [a]

Temperature	O_2 (mg/l)
0	14.16
5	12.37
10	10.92
15	9.76
20	8.84
25	8.11
30	7.53
35	7.04

[a] From Truesdale et al. (1955).

the oxygen saturation of the film is reduced to accept more oxygen from the atmosphere. Dissipation by molecular diffusion is almost negligible. Hutchinson (1957) quotes Grote on the rates of DO molecular diffusion in the water and gives the following example. Suppose a water body uniformly containing 11 mg/l oxygen comes into equilibrium with the atmosphere so that its surface layer contains 12.6 mg/l. After molecular diffusion had proceeded for 1 month, the concentration would be 12.2 mg/l at a depth of 3.1 cm, 11.8 mg at a 6.7 cm depth, and 11.4 at a 11.4 cm depth. From this it is obvious that DO dissipation is done mainly by eddy diffusion turbulence currents which are driven by the wind. Thus the wind has a considerable effect on the oxygen transfer rate through the interface. The wind can also increase the interface area, through which the transfer is done, by producing waves and air bubbles on the upper water surface.

DO concentration is thus affected by three main factors: (1) the rate of oxygen production through photosynthesis, (2) the rate of oxygen consumption by respiration, and (3) the rate of oxygen transfer through the air–water interface. When a pond is rich with phytoplankton, during the day photosynthesis may be intensive, and there is then considerable production of DO in excess of the requirements for respiration. As a result, DO is accumulated in the water, and saturation is usually reached or exceeded. Excess DO production can be lost to the atmosphere, but since the interface creates a barrier, very often the accumulation is faster than the emission into the atmosphere and supersaturation occurs. The transfer of DO across the interface is proportional to the saturation surplus (or deficit) of DO in the pond water. For constant wind conditions, there is usually a linear relationship between DO transfer across the interface and the percentage of oxygen saturation in the water (Odum, 1956). Thus the higher the supersaturation, the faster the DO emission until a balance is reached between DO production and emission. This balance can be reached at 200% saturation and more. The milder the wind, the higher the DO concentration at balance, while stronger winds increase oxygen emission and reduce DO concentration at balance.

During the night, photosynthesis and DO production cease, but not DO consumption. The DO concentration decreases until saturation is reached. When DO concentration is reduced below saturation, oxygen from the atmosphere is dissolved in the water. But here again, the interface is a barrier which reduces DO transfer. Very often consumption is faster than solution, and the DO concentration decreases until it reaches a minimum at daybreak. When the wind is strong this minimum is reached at a higher concentration, but when the wind is mild or still and consumption rapid, oxygen may be completely depleted, causing anoxia. A daily cycle in DO concentration is thus evident in productive ponds (Figure 11.1). The more productive the pond, with a larger population of phytoplankton and other organisms (such as in a fertilized pond), and the higher the temperature stimulating the photosynthesis and respiration processes, the stronger will be the daily fluctuation in DO concentration. From this DO cycle it is evident that the critical period during the day for anoxia is the early morning, just before sunrise.

11.1.2 Anoxia and Its Causes

Each of the main factors affecting the DO concentration and daily fluctuations in the water—production through photosynthesis, consumption by respiration, and the rate of DO transfer through the interface—is affected in turn by a large number of other biological and physical factors, some of which have multiple effects on DO. Some of the more important factors are:

Concentration of Phytoplankton Algae.   Though algae are DO producers, they also reduce light penetration and consume DO in the dark. Excess algae therefore reduce DO production in the lower layers of the water and increase the rate of depletion of DO during the night.

Heterotrophs.   The standing crop of heterotrophic organisms (bacteria, protozoa, zooplankton, fish, etc.) all consume oxygen both by day and by night.

Content of Organic Matter in the Water.   This immediately affects the bacterial population which increases rapidly and consumes oxygen.

Light.   Light is a prerequisite for photosynthesis and, therefore, DO production. Both surface light intensity and its penetration into the water play an important role here.

Temperature.   This can have a multiple effect. It determines the DO saturation of the water. The higher the temperature, the lower the DO concentration at saturation level. It also affects the rate of photosynthesis and the rate of respiration. In both cases an increase in temperature stimulates the process. It also affects the rate of transfer of O_2 through the atmosphere–water interface. The higher the temperature, the quicker is the transfer of DO through the interface.

Wind.   Wind is one of the most important climatic factors affecting the transfer of DO through the interface because of the eddy diffusion it creates. The stronger the wind, the higher the transfer rate.

Anoxia can result from different combinations of the factors mentioned above. It is not always possible to determine how anoxia is incurred

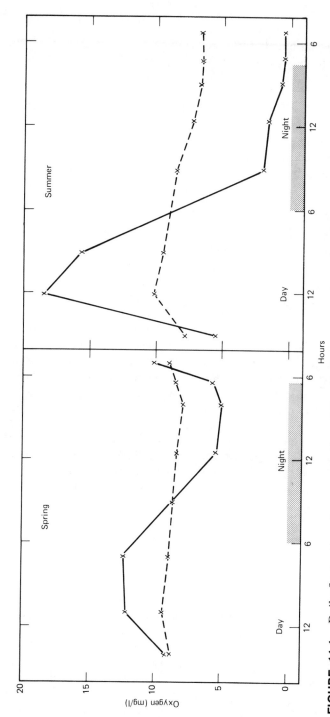

FIGURE 11.1. Daily fluctuations in concentration of dissolved oxygen in fertilized (solid line) and unfertilized (broken line) ponds at Dor, Israel, during two seasons of the year.

and what the combined factors responsible for it are. Four such combinations which seem to be common are given below. No doubt there are other possible combinations that may cause anoxia.

Excess Detritus. A high concentration of nonliving organic matter (detritus), which undergoes rapid decomposition in the water, can lead to anoxia. The organic matter can originate in the pond itself, as in the case of a sudden kill of plankton, or it can come from outside the pond. The kill of plankton can be caused by natural causes due to changes in environmental conditions. This has been observed and studied especially with blue-green algae which may develop to very high concentrations ("blooms") and then suddenly become pale and die due to photo-oxidative processes within the cells. According to Abeliovich (1972) this happens when growth conditions are not balanced, such as when the cells are strongly illuminated but CO_2 is insufficient for photosynthesis. The kill of plankton also can be caused by man. Such cases have already been discussed in relation to weed control (see Section 8.5). It can also happen when copper sulfate is added to control *Prymnesium parvum* (see Section 11.2). Organic matter can also come from outside, as in the case when excessive amounts of manure, wastewater, or any other wastes rich in organic matter (e.g., dairy or sugar industry wastes) are added to ponds. In all these cases the increase in organic detritus is followed by an increase in heterotrophic bacteria and the consumption of DO by them. Sometimes, during the day, enough DO is produced by photosynthesis to meet the increased demand, but during the night, especially if it is warm and calm, all DO reserves are depleted and fish die of anoxia before sunrise. More often, when the phytoplankton population in the pond dies, as in the case of blue-green algae mentioned above, photosynthesis is so weak that anoxia also occurs during the daytime.

Excess Phytoplankton. When light conditions are favorable, an increase in phytoplanktonic algae is associated with increased DO production. However, due to the self-shading effects of algae, light penetration into the deeper layers of the water is reduced and so is photosynthesis and DO production. Moreover, respiration which is not light dependent continues, and the DO balance in these deep layers may become negative, that is, consumption is greater than production. This is exacerbated when algae tend to float or produce a scum on the water such as with blue-greens. When the day is calm, DO concentration in the thin upper layer of pond water becomes very high, exceeding saturation. Due to a high DO gradient between water and atmosphere, the loss of oxygen through the interface is usually high. When the wind increases in the evening, the relatively thin upper layer, rich in DO, is mixed with a thick underlayer, poor in oxygen, and the resulting overall concentration, which is the reserve for nighttime respiration, is low. If the night is warm and calm, anoxia will often occur before sunrise.

Cloudiness. Cloudiness is a major cause for DO depletion and anoxia in some geographical areas during certain seasons of the year. Clouds reduce light irradiation considerably and thus photosynthesis and DO production. If winds are too mild to produce water currents and increase DO solution from the atmosphere, anoxia may occur. Cloudiness may have a severe effect when coupled with high DO consumption, such as described in the two previous cases.

Unbalanced Plankton Populations. At times zooplankton populations develop rapidly to high levels and consume much of the phytoplankton. These populations can be of rotifers or copepods and, when temperatures are low, also cladocerans. These zooplanktons consume a considerable portion of the oxygen and anoxia may occur. Conditions are sometimes worsened by a sudden kill of the zooplankton due to low oxygen and lack of food. Algae can then develop again. Such cycles between zooplankton and phytoplankton are quite common in wastewater oxidation ponds with relatively low organic loads. In fishponds, the fish prey heavily on zooplankton and stabilize conditions so that such cases seldom occur.

11.1.3 The Control of Anoxia

There are three different ways to control anoxia in ponds: biological, physical, and chemical.

Biological

Biological control seems to be the best way since it is prophylactic. It involves the choice of a proper combination of polyculture, as has already been discussed above (see Section 7.3.1). Silver carp consume excess minute algae; bighead carp reduce the concentration of zooplankton; and *Sarotherodon aureus* and mullets reduce the amount of detritus. These species must, of course, be stocked at appropriate ratios according to pond productivity and environmental conditions.

Physical

Aeration is the most common and immediate physical remedy for anoxia. This can be done by introducing air bubbles into the pond water by blowers through a system of perforated pipes. Oxygen passes

FIGURE 11.2. Aeration of a fishpond by enforced air bubbles.

from the bubbles into the water (Figure 11.2). In rearing ponds the pipe system usually occupies only a small part of the area. This part is intended to be a refuge for fish in times of anoxia. In relatively small storage ponds, the pipe system occupies a larger portion of the area.

The main drawback of this method is low efficiency. The bubbles stay in the water for only a short time, allowing only a little oxygen to dissolve. This is especially true for large bubbles. The smaller the bubbles, the larger is their total surface for a given volume of air and the lower their buoyancy. This increases the efficiency of oxygen transfer. For this purpose special pipes with very small pores are required, and these often clog with silt and algae.

Where no electrical connections are available, a water jet exhauster can be used. This device is basically a venturi-type aspirator in which water is introduced under a pressure of 2–2.5 atm through a jet of about 4 mm into a horizontal pipe which draws air through a vertical pipe which opens above the water surface.

A much more effective system involves creating turbulence in the water so that the transfer rate of oxygen through the atmosphere–water interface is increased. Schroeder (1975a) has suggested combining aeration and agitation. He examined the effect of aeration through a submerged vertical hollow tube, which has an inside diameter of 8 cm, is 0.5–1.0 m long, and is packed with plastic netting having 4 mm openings. As the air rises in the tube it acts as an "air lift pump," agitating the water. At the same time the bubbles are broken up when passing through the netting, and their rate of ascent is reduced. This was found to be much more efficient than simple aeration with no agitation.

The "Japanese water mill" is another device to agitate the water and create turbulence (Figure 11.3). It is composed of a 0.25–3 hp

FIGURE 11.3. Aeration with a Japanese water mill.

motor installed on a float. It is connected by two side axles to two paddle wheels. The wheels are submerged in the water to about one-third of their diameter. They rotate at a speed of about 120 rpm. They splash quite a lot of water, but their main advantage is the turbulence they create in the pond itself. Mitchell and Kirby (1976) and Rappaport et al. (1976) describe propeller-type devices which pick up water near the surface and discharge it against a cone to create a circular spray pattern and considerable surface turbulence. This, in principle, has an action similar to the Japanese water mill. Both groups, Mitchell and Kirby (1976) and Rappaport et al. (1976), found these devices more efficient than the application of air into the water. From experience it seems that one Japanese water mill of 3 hp is sufficient for a pond of about 3 ha.

Turbulence can also be achieved by pumping water out from the bottom layer of the pond and spraying it back onto its surface. In this way, in addition to the currents created, the surface area of the spray is considerably increased, thus increasing the oxygen absorption capacity of the water. The pump in this case can be installed on a small boat which can be transferred to any pond when needed.

One of the most common ways to remedy anoxia, especially when all the mechanical means mentioned above are not available, is the addition of water from outside sources to the pond in order to enrich it with oxygen. In severe cases, part of the water in the pond is exchanged. Since the total amount of oxygen thus added is usually not enough, and the dilution with pond water is large, the efficiency of this method is often limited.

In some cases a number of mechanical means are employed to enrich inflowing water with oxygen. These are usually devices like waterfalls, cascades, and sprinklers. They are effective only when the oxygen level of the inflowing water is lower than saturation. If the water is saturated or supersaturated, such means will either be of no use or will cause a reduction in oxygen. When the inflow water is cool and saturated, spreading it into a thin layer may cause its rapid warm-up and thus a loss of oxygen.

It is very important to remember that all these methods for aerating water increase oxygen absorption when its level is below saturation. When the water is supersaturated, these means will increase oxygen emittance from the water into the atmosphere. Aerators cannot be operated during the daytime as a prophylactic measure against anoxia the following night (unless the oxygen is also low during the day). This will worsen the situation since it will drive oxygen out and leave a smaller reserve for the night. Aerators should be operated only when a deficit develops, mostly during the night and early morning. In ponds densely stocked with fish, aerators can be operated automatically by a clock mechanism during critical hours.

Chemical

A rather controversial chemical treatment to remedy anoxia is the application of simple superphosphate at a rate of 120–150 kg/ha. This treatment has never been examined under controlled experimental conditions. Nevertheless, some fish farmers in Israel claim this treatment is beneficial against anoxia. Others could not see any improvement and consider it wasteful. The controversy may result from different sets of conditions causing the anoxia. The beneficial effect of superphosphate, if there is one, may be explained in the case of anoxia caused by a sudden natural death of phytoplankton (see Section 11.1.2). Two factors contribute to anoxia in this case: (1) the sudden increase in colloidal organic particles suspended in the water, which increases the bacterial population and the DO consumption; and (2) the reduction in DO production by photosynthesis.

Remember that simple superphosphate is a mixture of calcium sulfate (gypsum) and monocalcium phosphate. Gypsum is relatively more soluble than most of the common calcium salts. The Ca ion causes flocculation of colloidal organic particles by a physicochemical process. The organic flocs are precipitated to the bottom of the pond, relieving anoxia. At the same time the phosphate fraction stimulates the production of new algae populations and, thus, photosynthesis. These effects do not necessarily hold when a different set of conditions cause anoxia.

11.2 TOXIC ALGAE

A number of algae have been reported toxic to fish. Most of these belong to two groups: dinoflagellates and blue-greens.

Dinoflagellates such as the "red tide," a bloom of the algae *Gymnodinium breve,* have caused massive mortalities of fish in the Atlantic Ocean off the coast of Florida. It is estimated that 40–80 million tons of fish died in 1947 in that area. Matida et al. (1967) have reported a mortality of fish in Lake Sagami, Japan, caused by a bloom of another dinoflagellate, *Glenodinium gymnodinium.*

The second group is the blue-greens. Ingram and Prescott (1954) list a number of cases where mortalities of fish were caused by blooms of blue-green algae, mainly *Aphanizomenon flos-aquae* and *Microcystis flos-aquae* in freshwater lakes. Most of these cases occurred in natural bodies of water, however, and only a few were reported from fishponds, except in Israel. In Israel a toxic algae, the phytoflagellate *Prymnesium parvum,* causes considerable problems. When it first appeared in 1947, *Prymnesium* threatened the entire Israeli aquaculture industry. Sarig (1971) has estimated that it now affects an area of about 3000 ha, about 60% of the fishpond area, which causes an annual economic loss of about $100,000 in fish mortality and in the costs of control.

Prymnesium is a marine algae that has adapted to brackish waters. It does not grow at salinities below 0.1 parts per thousand (ppt). In practice it does not create problems in ponds having a salinity lower than 0.4 ppt. Under certain conditions it can rapidly develop into a bloom. The algae can secrete extracellular toxin, which causes the mortality of fish and some other organisms living in the water. Fish affected by the toxin behave in a specific way: they concentrate in the shallow parts of the pond where the toxin concentration is seemingly lower due to its absorption onto the silt, and they try to leap out of the water. Later they become sluggish and apathetic and, finally, die within 5–30 hours. Mortality within a pond is always total. Such mortalities have been observed in brackish lakes in The Netherlands and Denmark. *Prymnesium* appears in Israel mainly in the winter months (November to April) and only seldom in summer.

Reich and Aschner (1947) found that an ammonium sulfate concentration of 10 mg/l controls *Prymnesium,* and this fertilizer is often used. A careful study of *Prymnesium,* its toxins, and methods for its control was carried out by Shilo and his colleagues and has been quoted in detail by Sarig (1971). This study revealed a number of points that have modified the use of ammonium sulfate:

1 Since the specific environmental conditions responsible for the appearance of the *Prymnesium* bloom do not usually change with treatment by ammonium sulfate, *Prymnesium* often reappear after a short period. Control of *Prymnesium* can thus be expensive.

2 It was found that the growth of the phytoflagellate and toxin formation have different optimal requirements and that the toxin is easily inactivated. The concentration of toxin which is dangerous to fish is not necessarily correlated with the number of *Prymnesium* cells. It often happens that the concentration of *Prymnesium* cells is high, while that of the toxin is low and does not endanger fish.

3 The effect of ammonium sulfate on *Prymnesium* depends on pH and temperature. When the water temperature drops below 15°C and the pH below 8.5, much higher concentrations of ammonium sulfate are required to affect the *Prymnesium*.

These conclusions have led to a routine procedure for controlling *Prymnesium*, which was adopted in most fish farms in Israel. Water samples are taken from each pond, usually twice a week (the frequency is based on the season and occurrence of blooms), for microscopic analysis. If *Prymnesium* is found, a bioassay is conducted with the pond water, using *Gambusia* as a test fish. This bioassay, which is described in detail by Sarig (1971), is much more sensitive to the toxins than are the pond fish. It is only when the fish in the assay die that treatment is provided. The treatment varies with season according to pH and temperature. When the temperature is above 20°C, ammonium sulfate (10–17 ppm) is applied according to pH. The lower the pH, the more ammonium sulfate is applied. At pH below 8.5, liquid ammonia is more effective since it increases the pH by 0.4–1.0 pH units. At temperatures of 17–20°C, liquid ammonia applied at a rate of 10–14 ml/m³ is much more effective than ammonium sulfate. It is sprayed from a boat fitted with a closed plastic tank. Other equipment is the same as for fertilization with liquid ammonia (see Section 9.2). When water temperature is below 17°C and pH is below 8.5, both ammonium sulfate and liquid ammonia become ineffective. Copper sulfate can then be applied at a rate of 2–3 ppm for the control of *Prymnesium*.

As pointed out in Section 11.1.2, mass killing of pond algae may cause anoxia and fish mortality. This is especially dangerous when copper sulfate is used because it does not discriminate between the *Prymnesium* and other algae in the pond. Special care by aeration should therefore be given to this problem to prevent anoxia. On the other hand, the use of ammonia, whether as ammonium sulfate or liquid ammonia, creates a problem in nursing mullet fry or fingerlings.

These are very sensitive to ammonia, so ponds containing them should not be treated with ammonia. Copper sulfate is then the more desirable treatment chemical.

11.3 DISEASES AND PARASITES

11.3.1 Diseases

Bacterial and viral diseases are more common in temperate than subtropical and tropical climates. The reasons for this are not entirely clear, but they seem to relate to the fact that severe winters and the weakening of fish during that period because of harsh conditions predispose fish to attack by pathogenic bacteria and viruses. Interesting studies by Avtalion et al. (1970, 1973) have shown that at low temperatures, below 15°C, carp not only do not produce an immune response toward antigens, but rather develop a tolerance for them. If not immunized during the summer at high temperatures, the fish which have been in contact with pathogens during winter may thus be more susceptible to bacterial and viral diseases. Such processes may have an important role in the geographic distribution of fish diseases. The following paragraphs discuss the most common bacterial and viral diseases of warm-water fish in ponds.

Infectious Carp Dropsy. This is the most widespread disease of pond fish in Europe. At times it threatened to annihilate fish culture in the region. In the course of the infection the following symptoms are observed: hyperemia, which can be recognized by the red coloring, mainly of the abdomen; shallow irregular hemorrhages in the skin; and blisters in the skin filled with serumlike fluid. When these blisters break, lesions change into deep sores in the musculature. There are hemorrhages and necrotic areas in the gills, and hemorrhages in the internal organs and peritoneum, which can also be colored with bile pigments. The kidney and spleen are swollen and the gall bladder distended. The fish show skeletal malformation and severe anemia. The number of erythrocytes is sometimes reduced by about 50%. Affected fish are often excited and jump out of the water.

The etiology of the disease has been discussed by Otte (1963) and later by Schäperclaus (1965) who studied this disease for many years. Their conclusion is that the disease is caused primarily by the bacteria *Aeromonas punctata* (syn. *A. liquefaciens*) which may act with other bacteria such as *Pseudomonas fluorescens* as coagents. These bacteria are normally found in large numbers in fresh water, but they become violent when associated with a virus which seems to act as a provocating preparatory agent. Poor nutritional conditions due to harsh winters result in emaciated carp with very low fat reserves and low blood protein levels. Such fish are more susceptible to attack by disease agents.

The control of infection is mainly prophylactic. It is based on proper care of the ponds by periodic drying, fertilization to enhance production of natural food, stocking healthy fish for rearing, and feeding them properly so that they enter the next winter healthier and with larger fat reserves.

Antibiotics are usually injected into fish stocked in spring in the rearing ponds as a prophylactic measure. The recommended antibiotics are chloramphenicol (Chloromycetin), streptomycin, Aureomycin, and others. These antibiotics should be injected intraperitoneally in aqueous solution. Chloramphenicol, which does not dissolve easily in water, is dissolved first in a small amount of ethyl alcohol. The concentration of antibiotics should be such that the dosage per fish will be contained in 1 ml. This dosage is:

Chloromycetin   10–15 mg/kg live weight
Streptomycin   80–100 mg/kg live weight

Other antibiotics are given at the same dosage as the Chloromycetin. An overdose of streptomycin may be toxic.

Furunculosis.   In warmer climates, when fish are under conditions of stress, such as overcrowding, lack of oxygen, or low temperature, furunculosis may occur. This is characterized by necrosis and ulceration of the skin and degenerative changes in the kidneys and the liver. In severe cases the fins, especially the caudal fins, are eroded. The disease may be caused by the same bacteria that cause infectious carp dropsy in Europe, or more often by a related bacteria—*Aeromonas salmonicida.* If stress conditions persist, most of the fish may die, but with improvement of conditions, the ulcers heal rapidly and the fins regenerate. The best treatment is thus improvement of the environmental conditions.

11.3.2 Parasites

In warm climates the external parasites are the main group of pathogenic organisms causing mortality of fish ponds. A very good description of the most important parasites and ways to control them has been given by Sarig (1971). The reader is referred to his book for detailed information.

A most important contribution by Sarig and his colleagues was the development of practical ways to control parasites in commercial fishponds. Treatments suggested in earlier literature usually involved dipping infested fish for a short time (seconds or minutes) in a bath containing a relatively high concentration of disinfectant. The concentration of the disinfectant was at times so high that it killed some of the treated fish. Obviously such methods are not applicable to modern commercial aquaculture where tons of fish must be handled for treatment. Moreover, treating fish without removing the source of their infestation in the pond itself usually does not solve the problem.

The development of effective new low-concentration pesticides for parasites made it possible to treat fish with relatively small amounts of chemical added directly into ponds. The concentration of pesticide is brought to the desired effective level without removing fish from the ponds.

Sarig (1971) states that these factors characterize a pesticide suitable for use in fishponds:

1 There must be a wide safety margin between the concentration of the pesticide affecting the parasite and that lethal to fish. The ratio between these concentrations should be at least 1 : 5.

2 The pesticide should be water soluble, simple, and safe to handle.

3 It should not significantly lower the productivity of the pond.

4 The pesticide should be short-lived and easily degradable so as not to harm food organisms for long periods or be accumulated in fish flesh.

5 The cost of the pesticide must not be prohibitive.

Some of the parasites occuring in fishponds are especially dangerous to fry but are not usually a real hazard in later stages of life, when the fingerlings weigh over 3–5 g. The parasitic monogenean and digenean trematodes are good examples of such parasites. Adult fish may carry them endemically on their gills without much harm. These have already been discussed in Chapter 6.

Of the parasites which affect adult fish, some may be very dangerous and cause high mortalities in ponds. Others, however, do not actually kill fish, unless the parasites occur in very large numbers, but growth rate and market value of the fish may be reduced.

Some of the parasites occurring in rearing ponds are discussed briefly below. These usually belong to a number of diversified groups of organisms.

Ichthyophthirius multifilis is a protozoan ciliate that causes "white spot disease" and can be harmful to warm-water fish species. Heavy infestations usually occur during colder months, though sporadic outbreaks also occur in summer. The parasite attacks all the external parts of the fish, especially the gills where it hampers the exchange of gases (oxygen, CO_2, and ammonia). It thus interferes with normal respiration, excretion, and ionic balance, which seems to be the main reason for death. Fish recovering from moderate to heavy infestations have an increased resistance to reinfestation (Hines, 1972).

Ichthyophithirius can be controlled by spraying the pond with malachite green to bring about a concentration of 0.15 ppm. Three applications at 3 day intervals are required to eradicate the parasite completely. Care should be taken to use a pure malachite green since some products have a high content of zinc, which is toxic to fish.

Other protozoa such as *Trichodina* sp. (a peritrichous protozoan) and *Costia necatrix* (a flagellate) may infest fish skin and gills when the fish are crowded and weakened by anoxia or otherwise stressed. They may cause mortalities. Since these parasites are often found in association with other parasites such as *Ichthyophthirius*, it is difficult to determine their importance as pathogenic agents.

Two parasitic copepods quite common in fishponds in most parts of the world are fish lice, *Argulus* sp., and the anchor worm *Lernaea cyprinacea*. While they do not cause heavy mortalities in adult fish, they may harm them indirectly. They may also reduce the market value of the fish. They can be vectors for transferring bacterial and viral diseases, and they can facilitate infections by secondary parasites such as the fungus *Saprolegnia*.

Argulus is one of the most widespread freshwater ectoparasites in the world. *Argulus* seems to be a vector for transferring infectious carp dropsy and the viral disease epithelioma. *Argulus* is routinely found on carp in ponds. It attaches itself to the skin and fins by means of two suction discs and sucks the body fluids through a sharp proboscis. While they seldom cause mortality in fish, large numbers of *Argulus* may cause retardation of growth or weight loss.

Effective control can be achieved by applying a number of pesticides such as Lindane (γ-1,2,3,4,5,6-hexachlorocyclohexane) and Malathion [0,0-dimethyl S-(1,2-dicarboethoxyethyl) dithiophosphate]. The parasite has, however, shown increasing resistance to repeated treatments with Lindane, and within a few generations it may become immune to this pesticide. A much more effective pesticide, which has a larger safety margin for fish, is Bromex (dimethyl 1,2-dibromo-2,2 dichlorethyl phosphate) at a concentration of 0.2 ppm active ingredient in pond water. Some preparations of Bromex are emulsifiable concentrates containing only a certain percentage of the active ingredient. "Bromex-50" used in Israel contains about 42% (weight/volume) active ingredient. This should be taken into account when applying this pesticide.

While there are a number of *lernaea* species that affect fish, *L. cyprinacea* is the most abundant. It is an external parasite that attaches itself to the skin and fins of carp and other fish by means of special hooks in its head region. The head becomes embedded deep in the flesh of the fish within a distinctive lesion. Only the elongated wormlike body protrudes.

A few *L. cyprinacea* do not seem to affect fish much but make them aesthetically unacceptable as food and reduce their market value. Heavy infestations on small fish cause anemia and death. Knowledge of the life cycle of the parasite is important for its control. This has been studied by Lahav and Sarig (1964).

Only the female parasite reaches maturity on the fish. The female develops two egg sacs from which about 500–700 free-swimming nauplii are hatched. There are three stages of nauplii which, after 2–4 days (at 25–30°C), develop into the copepodid stage. At this stage they passively enter the gill cavity of the host fish and, upon reaching the fifth copepodid stage, they start crawling on the host's body surface. During this stage sexual differentiation occurs. The females are considerably longer (about 1.1 mm) than the males (0.8 mm). During the sixth and final stage the females are fertilized and penetrate the fish's skin and become absolute parasites. The parasites stimulate a strong tissue reaction at the point of penetration. The surrounding area becomes inflamed before the parasite can be seen with the naked eye. A red, dark lesion is formed in which the parasite becomes embedded. Fish recovering from moderate to heavy infestations have increased resistance.

The adult form of the parasite attached to the host is very hardy and resistant to pesticides. Even when affected, such as by $KMnO_4$, young parasites soon develop and appear on the fish 50–70 hours after treatment. On the other hand, the free-living larval stages are sensitive to insecticides. The life of these stages extends for 6–12 days at summer temperatures of 25–35°C and for 17 days at a temperature of 20°C. The application of Bromex, to make a concentration of 0.12–0.15 ppm active ingredient in pond water, kills these free-living larval stages (remember that commercial preparations contain 40–50% active ingredient). This application is repeated three times at intervals of 7 days in summer (28–32°C) and 12–14 days at lower temperatures (15–20°C).

A fungal infection which is abundant in fishponds is *Saprolegnia* sp. This ubiquitous parasite is normally present in water, but it attacks carp and other fish only as a secondary agent, usually in the winter months when they are weakened by low temperatures or are under stress following injury due to handling and other factors.

The mycelia project from the skin, fins, and gills and may cover large parts of the body. Fish losses caused by *Saprolegnia* may sometimes reach 50%. Expecially sensitive are tilapia when handled in marginal temperature conditions (10–14°C) when these prevail for some time (over 10 days). Gray mullet which have lost scales during handling and transportation are also usually attacked by this fungus.

Control can be achieved prophylactically by treating rearing ponds with formalin (technical grade). A concentration of 25 ppm in pond water should be applied immediately after stocking the fish (especially tilapia and mullet) in winter or when otherwise handling the fish. A series of three applications at 3 day intervals will clean the fish of *Saprolegnia*.

11.4 OFF-FLAVORS IN FISH

Though this problem cannot be considered hazardous to fish, it is a serious concern for the fish farmer since fish with an objectionable off-flavor—usually an earthy, musty taste—are rejected from the market and there is usually a subsequent drop in the acceptance of fish for a long period.

The occurrence of off-flavor in fish seems to be a phenomenon of universal distribution that is not related to specific climatic or geographic conditions. It was observed centuries ago in carp ponds in China, Japan, and Europe (Lovell, 1979) and recently also in Israel (Aschner et al., 1969), as well as in catfish ponds in the United States (Lovell, 1979).

The cause of the off-flavor was found to be a compound called *geosmin* produced by actinomycetes (Thaysen, 1936; Gerber and Lechevalier, 1965) and a number of blue-green algae of the genus *Oscillatoria*, such as *O. princeps, O. agardhi* (Cornelius and Bandt, 1933), *O. tenuis, O. prolifica, O. limosa* (Aschner et al. 1969), *O. chalybea* (Leventer and Eren, 1969), and *O. muscorum* (Lovell, 1979; Safferman et al., 1967). All these organisms grow on mud that is high in organic matter. They seem to prefer conditions in the transitory layer between the reduced mud and the oxidized water layer above it. These conditions are enhanced by the accumulation of organic matter in the bottom mud. The organic matter is decomposed, causing reduction of the mud.

Off-flavor in fish can be controlled in two ways: (1) control in the pond of the algae producing off-flavor, (2) removal of off-flavor from the fish after they have developed it. In ponds where the *Oscilatoria* is controlled, the off-flavor disappears. Such a control has been carried out successfully by treating the empty pond with $CuSO_4$ (0.1 kg/m^2) and restocking it after 3 days. A single t treatment was effective for more than 6 months.

A practical control method used by fish farmers in the Upper Gallilee, Israel (though its efficiency has never been proven experimentally), is to frequently fertilize the pond with nitrogen. The idea behind this treatment is to supply enough nitrogen for competing green algae to be able to dominate, thus decreasing the advantage of the blue-greens which can thrive on lower nitrogen concentrations. It may be that the abundant phytoplankton developed in this way shades light from the bottom-living *Oscillatoria*. Lovell (1979) suggests increasing turbidity or muddiness of pond water through mechanical agitation. It can be done with a bottom-feeding fish such as carp or *Sarotherodon aureus,* which will also prevent the *Oscillatoria* from developing (Leventer and Eren, 1969). Suspended clay particles also seem to absorb some of the off-flavor compounds (Lovell, 1979).

The chemical control of blue-green algae in catfish ponds is being practiced by some farmers in the United States through periodic application of 0.34 kg/ha of copper sulfate crystals to the water (Lovell, 1979). This is risky, however, since it can easily kill off all the algae in the pond which decompose rapidly and may cause subsequent anoxia and fish kills.

The off-flavor of fish which has already developed can be removed by holding them in a pond free of the organisms producing off-flavor, preferably a running-water pond, for about 7–14 days. According to Lovell (1979), the rate of improvements depends on water temperature. The higher the temperature, the shorter the period required for improvement.

Another source of off-flavors in fish, very often more persistent than that mentioned above, is industrial wastes. Their foul taste and odors are usually concentrated in the fat deposits of the fish's body. The higher the fat content, the more intensive and persistent the off-flavor.

The most important chemicals in industrial wastes that impart off-flavors to fish are phenols, tars, and mineral oils. Very low concentrations of these chemicals affect fish. This is especially true of chlorinated phenols such as *o*-chlorophenol and *p*-chlorophenol, which impart to carp a distinct flavor when in concentrations of 0.015 and 0.06 mg/l, respectively. Eels and oysters are even more sensitive and get the off-flavor when the water contains as little as 0.001 mg/l

O-chlorophenol (Mann, 1969). A concentration of 5–14 mg/l mineral oil, or less if in suspension, also imparts a distinctive flavor. The flavor imparted by both phenols and oils becomes more intensive and is noticeable at lower concentrations when these become mixed with detergents.

Pollution with the chemicals mentioned above may occur when wastewater from refineries, or more often from garages or even water flushing city streets, is introduced into ponds. In some cases the off-flavor has been imparted while transporting the fish to market because of a leak in the mechanical devices of the transporting tank, especially the aeration blower and its engine. Since the off-flavor can only be remedied by holding the fish in fresh water for a rather long time, care should be taken to avoid these sources of pollution.

11.5 TOXICANTS

One of the most severe hazards in aquaculture is the poisoning of fish by chemical toxicants that somehow find their way into the ponds and kill the fish and/or their natural food. Some of these, such as heavy metal compounds, industrial wastes, and detergents, are usually associated with wastewater (see Section 9.5). Their effect must be taken into account when wastewater is used in a pond, and a proper monitoring system to detect their presence must be established.

Other kinds of toxicants, which may cause considerable damage in aquaculture, but which are not under the control of the fish farmer, are agricultural pesticides and herbicides. Some of these are extremely toxic to fish and may cause mortality at very low concentrations. According to Henderson et al. (1959), the chlorinated hydrocarbons [except BHC (Lindane)] may kill fish at concentrations as low as 0.6 ppb (Endrin on bluegill), which is equal to the application of 6 g of active ingredient to 1 ha of pond 1 m deep. The agricultural pesticides and herbicides are commonly sprayed from the air and often, when sprayed carelessly over ponds, may cause heavy mortalities of fish. Additional consequences are the possible effect of sublethal concentrations of the chemicals on the growth and yield of fish and the accumulation of the organic residues in fish flesh. Lakota et al. (1978) found anatomic-pathological changes in carp fry subjected to treatment with sublethal concentrations (0.025 ppm) of methoxychlor, and also an accumulation of this compound in the body itself of up to 1.8–11.14 ppm. Bridges et al. (1963) found 3–4 ppm of DDT and its metabolites in the flesh of fish from a farm pond 1 month after the application of 0.2 ppm DDT to the pond.

TABLE 11.2 Acute Toxicity Level for Some Pesticides and Herbicides for Pond Fishes

Compound Fish Species	96 hr LC$_{50}$[a] (ppm)	Reference
Chlorinated Hydrocarbons		
Endrin (1,2,3,4,10,10-hexachloro-6,7-epoxy-1,4,4a,5,7,8,8a-octahydro-1,4-endo-endo-5,8-dimethanonaphthalene)		
Channel catfish	0.0001–0.0002	Mount & Putnicki, 1966 (quoted by Johnson, 1969).
Tilapia melanopleura	0.0008	Gruber, 1959
Bluegill	0.0006	Henderson et al., 1959
Goldfish	0.002	Henderson et al., 1959
Aldrin (1,2,3,4,10,10-hexachloro-1,4,4a,5,8,8a-hexahydro-1,4-endo-exo-5,8-dimenthanonaphthalene)		
Bluegill	0.013	Henderson et al., 1959
Goldfish	0.028	Henderson et al., 1959
Common carp	0.035	Kimura and Matida, 1966
Dieldrin (1,2,3,4,10,10-hexachloro-6,7-epoxy-1,4,4a,5,6,7,8,8a-octahydro-1,4-endo-exo-5,8-dimethanonaphthalene)		
Bluegill	0.008	Henderson et al., 1959
Goldfish	0.037	Henderson et al., 1959
Common carp	0.03	Kimura and Matida, 1966
Chloredane (1,2,4,5,6,7,8,8-octachloro-2,3,3a,4,7,7a-hexahydro-4,7-methanoindene)		
Bluegill	0.022	Henderson et al., 1959
Goldfish	0.082	Henderson et al., 1959
Heptachlor (1,4,5,6,7,8,8-heptachloro-3a,4,7,7a-tetrahydro-4,7-methanoindene)		
Bluegill	0.019	Henderson et al., 1959
Goldfish	0.23	Henderson et al., 1959

TABLE 11.2 (Continued)

Compound Fish Species	96 hr LC$_{50}$[a] (ppm)	Reference
Toxaphene (octachloro camphene)		
Bluegill	0.004	Henderson et al., 1959
Goldfish	0.006	Henderson et al., 1959
DDT [2,2 bis-(p-chlorophenyl)1,1,1-trichloroethane]		
Tilapia melanopleura	0.05	Gruber, 1959
Bluegill	0.016	Henderson et al., 1959
Goldfish	0.027	Henderson et al., 1959
Lindane (gamma-1,2,3,4,5,6-hexachlorocyclohexane)		
Bluegill	0.077	Henderson et al., 1959
Goldfish	0.152	Henderson et al., 1959
Methoxychlor [1,1,1-trichloro-2,2-bis (p-methoxyphenyl) ethane]		
Carp fry	0.056	Lakota et al., 1978
Bluegill	0.062	Henderson et al., 1959
Goldfish	0.056	Henderson et al., 1959
Organic Phosphates		
Diazinon [O,O diethyl O-(2-isopropyl-6-methyl-4-pyrimidinyl) thiophosphate]		
Common carp	1.6	Kimura and Matida, 1966
EPN (O-ethyl-o-p-nitrophenyl benzenethiophosphonate)		
Fathead minnow	0.2	Henderson et al., 1960
Malathion [O,O-dimethyl S-(1,2-dicarboethoxyethyl) dithiophosphate]		
Common carp	3.8	Kimura and Matida, 1966
Methyl parathion (O,O-dimethyl-O-p-nitrophenyl tiophosphate)		
Common carp	5.1	Kimura and Matida, 1966
Fathead minnow	7.5	Henderson et al., 1960

TABLE 11.2 (Continued)

Compound	Fish Species	96 hr LC_{50}[a] (ppm)	Reference
Nankor (O,O-dimethyl O-2,4,5 trichlorophenyl phosphorothioate)			
	Common carp	0.83	Kimura and Matida, 1966
Parathion			
	Common carp	1.0	Lahav and Sarig, 1969
	Tilapia	0.85	Lahav and Sarig, 1969
	Mullet	0.5	Lahav and Sarig 1969
Cotnion (azinophosmethyl)			
	Common carp	0.14	Lahav and Sarig, 1969
	Tilapia	0.02	Lahav and Sarig, 1969
	Mullet	0.08	Lahav and Sarig, 1969
Inorganic herbicides			
Sodium arsenite			
	Bluegill	35.0	Gilderhus, 1966
	Goldfish	34.0	Gilderhus, 1966
Organic herbicides			
Baron (dalapon ester or 2,4,5-trichlorophenoxy ethanol)			
	Largemouth bass	4.6	Bond et al., 1960
Endothal (disodium 3,6-endoxyhexahydrophthalte)			
	Bluegill	2.5	Jones, 1965
Kuron [α (2,4,5-trichlorophenoxy) propionic acid]			
	Largemouth bass	3.5	Bond et al., 1960
	Channel catfish fry	0.9	Jones, 1965
	Bluegill fry	0.4	Jones, 1965

[a] Lethal concentration to 50% of fish exposed to the compound for 96 hours.

The toxicity of pesticides and herbicides will differ with different species and different sizes of fish, with temperature, and with different water composition (e.g., hard and soft water). In some cases resistance to the chemicals may develop if fish are exposed to sublethal concentrations for some time. There is therefore some difficulty in presenting a general chart of their toxicity. However, an attempt to compose such a chart from literature has been made in Table 11.2. Deviations from the figures given in this table may occur due to existing conditions.

The reaction of fish to most toxicants is similar. The reaction time may be different with different concentrations, chemicals, and species. At high concentrations, the first visible effects occur after 10–60 minutes. The fish at first show a brief period of high excitability, followed by alternate periods of muscular spasms, and then by a complete loss of equilibrium, with the fish turning on their longer axis. In practice not much can be done to save these fish. In aquaria and tanks, the replacement of contaminated water may save the fish, but this cannot be done rapidly enough in ponds.

12 Economics of the Fish Farm

The major interest of the fish farmer obviously lies in the profitability of his farm rather than in its production per se. This naturally involves two additional factors besides yield: returns from the sale of fish and the cost to produce them. Careful economic analysis of profitability should be made at two different stages in farm management: (1) before establishing the farm in order to determine whether the investment is likely to be profitable and at what level of intensity the farm must operate in order to be profitable; (2) during the continuing management of the farm to test whether, in fact, the operation is profitable and to examine ways of improving profitability.

In order to allow true comparisons with annual income, capital investment likewise should be expressed on an annual basis, together with any other costs incurred in the production process (e.g., wages, feed costs, etc.). This is done by calculating annual depreciation of the assets and annual interest on the capital invested to obtain them.

Capital investment per unit pond area will vary among countries and even in regions of the same country due to differences in the size of farms, the size of individual ponds, the equipment used to construct ponds, the costs of construction, and the costs of equipment. It is therefore risky to attempt generalized comparisons between regions or even farms, and each case should be analyzed individually. To illustrate this, the following data are presented on investment for construction and equipment (in U.S. dollars, not including the value of the land) in three different countries.

Hong Kong. These data are for polyculture of gray mullet and Chinese carps on a 1 ha farm in 1972 (from Pillay, 1973).

	Per Farm	Per Hectare
Construction	$19,000	$19,000
Equipment	4,011	4,011
	$23,011	$23,011

United States.   Here are data for channel catfish farming on a 64.8 ha
farm in Mississippi (Mitchel and Usry, 1967, quoted from Pillay, 1973).

	Per Farm	Per Hectare
Construction	$50,830	$ 784
Equipment	22,190	342
	$73,020	$1,126

Israel   These data are for polyculture of common carp, tilapia, Chinese
carp, and mullet on a 100 ha farm in 1978 (personal communication,
Israel Ministry of Agriculture, Agricultural Planning and Development Center).

	Per Farm	Per Hectare
Construction	$326,800	$3,268
Equipment	110,457	1,105
	$437,257	$4,373

From the preceding data one can see that the differences in investment
can be very large.

Table 12.1 gives a detailed breakdown of capital investment on a
100 ha fish farm in Israel. The depreciation times for ponds and equipment and the annual depreciation allowances are also included. The
annual depreciation allowance takes into account an 8% interest on
the capital.

The analysis of production operations on the farm is based on
records taken during these operations. On the basis of these records
new ways are sought to increase profitability. In some cases, where the
area-related (fixed) costs are relatively high and the yield-related (variable) costs are relatively low, intensification may result in higher
profits. This involves increasing the stocking density and the inputs to
sustain larger numbers of fish and their growth. The area-related costs
increase to some extent (more water, aeration, etc.), while yield-related
costs increase much more, especially due to increased costs for feed.
Economic limits to intensification occur when the marginal addition of
yield-related costs is equal to or higher than the value of the extra
yield gained by it.

Table 12.2 gives an economic analysis of a polyculture system at
two levels of intensification at the Beithshan Valley area, Israel (Mr. I.

TABLE 12.1 Investments for Construction and Equipment on a 100 ha Fish Farm in Israel (December 1977, in U.S. Dollars)[a]

Item	Units	Cost per Unit	Total Costs	Cost per ha	Depreciation Time (yr)	Annual Depreciation Allowance[b] per ha	Percent of Total
Construction	100 ha	3,268	326,800	3268	20	333	68.5
Truck	1	16,340	16,340	163	10	24	5.0
Jeep	1	5,229	5,229	52	10	8	1.7
Wheel tractors	2	5,229	10,454	105	10	16	3.3
Marketing tanks	1	3,268	3,268	33	10	5	1.0
Transport tanks	3	1,961	5,883	59	10	9	1.9
Tank for fry	1	980	980	10	10	2	0.4
Elevators	1	2,288	2,288	23	7	4	0.8
Small elevators	2	1,307	2,614	26	7	5	1.0
Ammonia storage tank	1	1,307	1,307	13	7	3	0.6
Ammonia distribution tank	1	980	980	10	7	2	0.4
Net drum	1	980	980	10	10	1	0.2
Nets	8	327	2,616	26	10	4	0.8
Lab. equipment		1,307	1,307	13	10	2	0.4
Demand feeders	20	980	19,610	196	10	29	6.0
Feeding blower tank	1	3,921	3,921	39	10	6	1.2
Storage and Packing Sheds		32,680	32,680	327	20	33	6.8
			437,257	4373		486	100.0

[a] From personal communication, Israel Ministry of Agriculture, Agricultural Planning and Development Center.
[b] Includes 8% interest on invested capital.
[c] Includes water system, drainage, piping, monks, etc.

233

TABLE 12.2 Economic Analysis of Polyculture at Two Levels of Intensification at Beithshan Valley, Israel (in U.S. Dollars)

a. Yield-Rated (Variable) Cost (Per Ton of Fish)

Item	C. Carp Lower Density	C. Carp Higher Density	Tilapia	S. Carp	Mullet
Marketing (4.5%)	108	108	246	168	230
Harvesting and storing	67	67	67	33	33
Fry	21	26	450	25	167
Feed	417	550	167	—	—
Interest on working capital (3%)	17	20	20	2	6
	630	771	950	228	436

b. Net Returns per Ton

Price per ton	1217	1217	1500	833	2167
Net return	587	446	550	545	1731

c. Net Return per Hectare

	Lower Density			Higher Density		
Species	Yield (ton/ha)	Return per Ton	Total Return	Yield (ton/ha)	Return per Ton	Total Return
C. carp	2.0	587	1174	4.0	446	1784
Tilapia	0.45	550	248	0.9	550	495
S. carp	0.45	545	245	0.9	545	490
Mullet	0.15	1731	260	0.3	1731	519
Totals	3.05		1927	6.1		3286

d. Area-related (Fixed) Costs (per hectare)

Item	Lower Density		Higher Density	
	Amount	Cost	Amount	Cost
Labor	20 days × 16.5	330	30 days × 16.5	495
Water	35,000 m³ × 0.01	350	35,000 m³ × 0.01	350
Water recycling		—	23,000 m³ × 0.005	115
Fertilizers	2 tons × 79	158	2 tons × 79	158
Manure	10 m³ × 5	50	15 m³ × 5	75
Pesticides	—	50	—	75
Transport		117		200
Various small equipment		83		100
Electricity		—	3000 kW × 0.03	90
Maintenance and repairs		250		250
Investment cost		486		555
Interest on working capital (3%)		6		8
Totals		1880		2471
e. Net Profit per Hectare		47		815

Peleg, personal communication). The lower intensification level had a lower stocking rate and produced 3.0 ton/ha at a feed conversion rate of 1:2.5. The higher intensification level produced 6.1 ton/ha with higher water circulation, aeration, and a feed conversion ratio of 1:3. In this particular case fish farming at the lower intensification level, which yields about 3 ton/ha, was not profitable. Only at higher intensification levels and higher yields did it become profitable.

The biological limits on intensification are different for each of the species in polyculture. It follows, therefore, that with increasing intensification, the species combination will also vary. Economic considerations will affect this variation as well by tending to increase production of the species which bring the highest income. Thus in the previous example (Table 12.2, part b), it can be seen that the return from 1 kg of mullet is much higher than from any other fish species in polyculture. It will be worthwhile to culture more mullet in polyculture even if, biologically, their presence reduces the production of other fish to some extent.

It would seem that an experienced fish farmer would soon find the most profitable species combination and density ("culture program"). A number of questions arise, however, which cannot be answered easily by the farmer. For instance: Where is the limit to intensification when the extra inputs become more expensive than the value of extra production? Also, what is the value of the production of one species which is lost due to competition when one increases the density and production of another species, and what are the economic results? Finally, what should be the optimal culture program for a given set of conditions and inputs?

The answers to these and other questions depend on many parameters. They can be answered only by defining the growth rate of each fish species in polyculture as a function of its density, the amount and quality of supplemental feed, composition of the water, water temperature, and the rate of water flow. To make it even more complicated, there is a strong interaction among growth rates, as defined above, of different species in polyculture. It seems that only a computer simulation model will solve these relations and supply a plan for a bioeconomically optimal system. Though some attempts are being made in this direction, it will be some time before such a model is available. In the meantime it can be done only by trial and error by the fish farmer, with modifications based on analysis of the economic results at the end of each culture period.

Moreover, even when the best system is found for a single pond, a number of constraints limit its application to the whole farm. A culture program which is suitable for a productive pond does not necessarily suit the unproductive ponds. The length of the season is also not always under control of the farmer. However, the most important constraint in this respect are market limitations and fry supply. Market limitations, if they exist, may be imposed as marketing quotas, as is the case in Israel, where the Fish Breeders' Association regulates the market. Market limitations also can be imposed by the marketing potential, or by the handling capacity of the farm itself.

The fry supply may become an important issue at higher intensification levels since the number of nursed fry is much higher and more ponds must be devoted to spawning and nursing. More storage ponds are also usually required to regulate marketing. Because of these constraints, the planning of fish culture becomes quite complicated. No doubt a computer model would be of considerable help here, but since such a model does not yet exist, it still remains in the hands of the fish farmer to analyze the economic results of the previous culture program and plan accordingly for the future. Auxiliary ponds (spawning, nursing, and storage) may be less profitable than intensive

TABLE 12.3 Breakdown of Operating Costs into Major Items as a Percentage of Total Costs in Six Different Farm Systems[a]

| Item | Fish Farming Systems[b] | | | | | |
	A	B	C	D	E	F
Stocking material (fry)	1.7	44.2	7.6	37.5	27.5	7.5
Feeds	39.4	0.3	39.0	31.9	36.6	24.3
Fertilizers	4.1	5.5	3.6	13.3	1.9	5.6
Labor	26.8	32.9	13.1	10.3	15.3	8.8
Pesticides	—	—	—	—	5.0	1.3
Maintenance and repairs	7.2	0.3	13.0	6.3	—	6.7
Water	—	—	—	—	—	9.5
Other costs	9.8	16.8	3.7	0.7	4.6	23.3
Investment costs	11.0	—	20.3	—	9.3	13.0
Totals	100.0	100.0	100.0	100.0	100.0	100.0

[a] Data for farms A to E are from Pillay (1973).
[b] A: Common carp culture in a 555 ha state farm in Poland, 1969. B: Culture of Indian carp in a 3.5 ha freshwater pond in West Bengal, India, 1971. C: Polyculture of grey mullet and Chinese carp in a 1 ha farm in Hong Kong, 1972. D: Milkfish culture in a 5 ha farm in Bulacan, the Philippines, 1972. E: Channel catfish farming in a 64.8 ha farm in Mississippi, 1967. F: Polyculture of common carp, Chinese carp, tilapia, and mullet in a 100 ha farm in Israel, 1977.

rearing ponds, but they are essential for the operation of the farm. Therefore they must also be taken into account when evaluating the economic merits of a culture program.

One way to increase profits is by reducing operating costs as much as possible. This can be done in biological and technological ways, such as reducing fry loss, labor costs, and other costs. Such reductions will have greater consequence when the cost of the item saved makes up a large part of the total costs. The breakdown of operating costs is different in different culture systems, and therefore the effort to reduce costs must be aimed at different targets. Table 12.3 gives a breakdown of the major items of operating costs as percentages of total costs for six different farm systems (A–F). It is obvious that costs for fry and fingerlings do not play an important role in Poland (A) or Hong Kong (C), but they are very important in India (B), and the Philippines (D), and efforts should be made to reduce them. Feed, on the other hand, constitutes a major item in the costs in Poland (A), and Hong Kong (C), and savings there will be of greater consequence.

Our discussion in this chapter should make it clear that aquaculture has become a modern agricultural production system. As such, and in order to bring expected profits, it must be managed with modern methods based on sound scientific, ecological, technological, and economic principles. We hope that this book contributes to man's knowledge of these principles and through that knowledge to the development of aquaculture in the world.

References

Abeliovich, A. 1972. Oxygen regime in fish ponds. Ph.D. Thesis. The Hebrew University, Jerusalem: vi + 88 pp.

Alabaster, J. S., and R. Lloyd. 1980. Water quality criteria for freshwater fish. Butterworths, London and Boston: 297 pp.

Alikunhi, K. H. 1966. Synopsis of biological data on common carp *Cyprinus carpio* (Linnaeus), 1958 (Asia and the Far East). FAO Fish Synop., Vol. 31.1: 70 pp.

Allanson, B. R., A. Bok, and N. I. Van Nyk. 1971. The influence of exposure to low temperature on *Tilapia mossambica* (Peters) (Cichlidae) II. Changes in serum osmolarity, sodium and chloride ion concentrations. J. Fish. Biol., 3(2): 181–185.

Appelbaum, S. 1977. Geeigneter Ersatz für Lebendnahrung von Karpfenbrut? Arch. Fischereiwiss., 28(1): 31–43.

Appelbaum, S., and U. Dor. 1978. Ten day experimental nursing of carp (*Cyprinus carpio* L.) larvae with dry feed. Bamidgeh, 30(3): 88–95.

Aquaculture Development and Coordination Programme. 1976. Aquaculture planning in Asia (ADCP/REP/76/2). United Nations Development Programme, FAO, Rome: 154 pp.

Aschner, M., C. Laventer, and I. Chorin-Kirsch. 1969. Off flavor in carp from fish ponds in the coastal plain and the Galil. Bamidgeh, 19(1): 23–25.

Avault, J. W., and E. W. Shell. 1968. Preliminary studies with the hybrid tilapia *Tilapia nilotica* × *Tilapia mossambica*. FAO Fish. Rep., 44(4): 237–242.

Avault, J. W., E. W. Shell, and R. O. Smitherman. 1968. Procedures for overwintering tilapia. FAO Fish. Rep., 44(4): 343–345.

Avtalion, R. R., and I. S. Hammerman. 1978. Sex determination in *Sarotherodon* (tilapia). 1. Introduction to a theory of autosomal influence. Bamidgeh, 30(4): 110–115.

Avtalion, R. R., Z. Malik, E. Lefler, and E. Katz. 1970. Temperature effect on immune resistance of fish to pathogens. Bamidgeh, 22(2): 33–38.

Avtalion, R. R., A. Wojdani, Z. Malik, R. Shahrabani, and M. Duczyminer. 1973. Influence of environmental temperature on the immune response in fish. Curr. Top. Microbiol. Immunol., 61: 1–35.

Backiel, T., and K. Stegman. 1968. Temperature and yield in carp ponds. FAO Fish. Rep., 44(4): 334–342.

Bakos, J. 1979. Crossbreeding Hungarian races of common carp to develop more productive hybrids. *In* T. V. R. Pillay and Wm. A. Dill, Eds., Advances in Aquaculture. Fishing News Books, Farnham, Surrey, England: pp. 633–3635.

Balon, E. K. 1974. Domestication of the carp *Cyprinus carpio* L. Life Sciences Miscellaneous Publication, Royal Ontario Museum, Toronto, Canada: 37 pp.

Bardach, J. E., J. H. Ryther, and W. O. McLarney. 1972. Aquaculture, the farming and husbandry of freshwater and marine organisms, Wiley-Interscience, N.Y.

Barthelmes, D. 1975. Elements der Sauerstoffbilanz in Karpfenteichen, ihr Wirkungsweise sowie die Optimierungsmoglichkeiten durch Silberkarpfen (*Hypophthalmichthys molitrix*). Z. Binnenfisch. DDR, 22(11): 325–333; (12): 355–363.

Blackburn, R. D. 1968. Weed control in fish ponds in the United States. FAO Fish. Rep., 44(5): 1–17.

Blackburn, R. D. 1974. Chemical control. *In* D. S. Mitchell, Ed., Aquatic vegetation and its use and control. UNESCO, Paris: pp. 85–98.

Bond, C. E., R. H. Lewis, and J. L. Fryer. 1960. Toxicity of various herbicidal material to fishes. *In* C. M. Tarzwell, Comp. Biological problems in water pollution 2. U.S. Dept. of Health, Education and Welfare, Robert A. Taft Sanit. Eng. Center, Cincinnati, Ohio: pp. 96–101.

Boyd, C. E. 1979. Lime requirement and application in fish ponds. *In* T. V. R. Pillay and Wm. A. Dill, Eds., Advances in aquaculture. Fishing News Books, Farnham, Surrey, England: pp. 120–122.

Bridges W. R., B. J. Kallman, and A. K. Andrews. 1963. Persistence of DDT and its metabolites in a farm pond. Trans. Am. Fish. Soc., 92(4): 421–431.

Brown, E. E. 1969. The fresh water cultured fish industry of Japan. Univ. of Georgia, College of Agric. Exp. Sta., Res. Rep., 41: 57 pp.

Buck, D. H., R. J. Baur, and C. R. Rose. 1979. Experiments in recycling swine manure in fishponds. *In* T. V. R. Pillay and Wm. A. Dill, Eds., Advances in aquaculture. Fishing News Books, Farnham, Surrey, England: pp. 489–492.

Chao, N. H., H. P. Chen, and I. C. Liao. 1974. Study on cryogenic preservation of grey mullet sperm. Aquiculture 2(2): 25–39.

Chen, F. Y. 1969. Preliminary studies on the sex determining mechanism of *Tilapia mossambica* Peters and *T. hornorum* Trewavas. Verh. Int. Ver. Limnol. 17: 719–724.

Chervinski, J. 1974. Sea bass, *Dicentrarchus labrax* Linne (Pisces, Serranidae), a "police-fish" in freshwater ponds and its adaptability to various saline conditions. Bamidgeh, 26(4): 110–113.

Chervinski, J. 1975. Sea basses, *Dicentrarchus labrax* (Linne) and *D. punctatus* (Bloch) (Pisces, Serranidae), a control fish in freshwater. Aquaculture, 6: 249–256.

Chervinski, J., and M. Lahav. 1976. The effect of exposure to low temperature on fingerlings of local tilapia (*Tilapia aurea*) (Steindachner) and imported tilapia (*Tilapia vulcani*) (Trewavas) and *Tilapia nilotica* (Linne) in Israel. Bamidgeh, 28(1/2): 25–29.

Chiba, K. 1965. Studies on the carp culture in running water pond—I. Fish production and its environmental conditions in a certain fish farm in Gumma Prefecture. Bull. Freshw. Fish. Res. Lab. (Tokyo) 15(1): 13–33.

Chimits, P. 1955. Tilapia and its culture: A preliminary bibliography. FAO Fish. Bull., 8(1): 1–33.

REFERENCES

Chimits, P. 1957. The tilapias and their culture. FAO Fish. Bull., 10(1): 1–24.

Collins, R. A., and M. N. Delmendo. 1979. Comparative economics of aquaculture in cages, raceways and enclosures. In T. V. R. Pillay and Wm. A. Dill, Eds., Advances in aquaculture. Fishing News Books, Farnham, Surrey, England: pp. 472–477.

Coon, K. L., A. Larsen, and J. E. Ellis. 1968. Mechanised haul seine for use in farm ponds. Fish. Ind. Res., 4(2): 91–108.

Cornelius, W. O., and H. J. Bandt. 1933. Fischereishaedigungen durch Starke Vermehrung gewisser pflanzlicher plankton insbesondere Geschmacks-Beeinflussung der Fische durch Oscillatorien. Z. Fisch., 31: 675.

Cowey, C. B., and J. R. Sargent. 1972. Fish nutrition. Adv. Mar. Biol., 10: 383–393.

Demoll, R. 1926. Die Reiningung von Abwassern in Fischteichen. Handb. Binnenfisch. Mitteleur., 6(2): 222–262.

Demoll, R. 1940a. Kartoffelpülpe als Karpfenfutter. Allg. Fisch. Ztg., 23: 2 pp.

Demoll, R. 1940b. Vorlaufige Mitteilung über die Kartoffel-Fütterungs—Versuche des Jahres 1939 in Wielenbach. Fisch.-Ztg., 43(8): 1 p.

Denzer, H. W. 1968. Studies on the physiology of young tilapia. FAO Fish. Rep., 44(4): 358–366.

De Silva, S. S., and M. J. S. Wijeyaratne. 1976. Studies on the biology of young grey mullet Mugil cephalus L. FAO Tech. Conf. Aquaculture, Kyoto. FIR:AQ/Conf/76/E. 34: 12 pp.

Doron, D. 1976. Harvesting, withdrawal and sorting of fish with new equipment and working systems (in Hebrew). Israel Ministry of Agric. Extension Service, Dept. of Technology: 13 pp.

Dunseth, D. R., and D. R. Bayne. 1978. Recruitment control and production of Tilapia aurea (Steindachner) with the predator Cichlasoma managuense (Günther). Aquaculture, 14(4): 383–390.

Eckstein, B., and M. Spira. 1965. Effect of sex hormones on the gonadal differentiation in a cichlid, Tilapia aurea. Biol. Bull., 129: 482–489.

Elster, H., and H. Mann. 1950. Experimentelle Beiträge zur Kenntnis der Physiologie der Befruchtung bei Fischen. Arch. Fischereiwiss., 2: 49–72.

Erokhina, L. 1959. Experiments on the use of pelleted feed (in Russian). Rybowod. Rybolow., 2(4): 8–10.

FAO. 1967. Report to the Government of Uganda on experimental fish culture project in Uganda, 1965–1966. Based on the work of Yoel Pruginin, FAO/UNDP(TA) Inland Fishery Biologist (Fish Culture). Rep. FAO/UNDP(TA), (2446): 16 pp.

FAO. 1978. 1977 Yearbood of fishery statistics. Catches and landings, Vol. 44. FAO, Rome: 343 pp.

FAO/UN. 1965. Report to the Government of Uganda on the experimental fish culture project in Uganda, 1962–1964. Based on the work of Yoel Pruginin. Rep. FAO/ EPTA:i + 25 pp.

Fijan, N., A. Jeromel, V. Krizanac, and E. Teskeredzic. 1976. The floating bag—A simple demand feeder for fish. Rib. Ital. Piscic. Ittiop., 11(1): 22–23.

Fish, G. R. 1955. The food of Tilapia in East Africa. Uganda J., 19: 85–89.

Fishelson, L. 1962. Hybrids of two species of fishes of the genus *Tilapia* (Cichlidae Teleostei) (in Hebrew with English summ.). Fishermen's Bulletin (Haifa) 4(2): 14–19.

Fishelson, L. 1966. Cichlidae of the genus Tilapia in Israel. Bamidgeh, 18(3–4): 67–80.

Fleming, R. H. 1940. Composition of plankton and units for reporting populations and productions. Proc. Pac. Sci. Congr., 6(3): 535–540.

Fryer, G., and T. D. Iles. 1972. The cichlid fishes of the great lakes of Africa, their biology and evolution. Oliver and Boyd, Edinburgh: 641 pp.

Gaudet, J. J. 1974. The normal role of vegetation in water. *In* D. S. Mitchell, Ed., Aquatic vegetation and its use and control. Unesco, Paris: pp. 24–37.

Gerber, N. N., and H. A. Lechevalier. 1965. Geosmin, an earthy-smelling substance isolated from actinomycetes. Appl. Microbiol., 13: 935.

Gilderhus, P. A. 1966. Some effects of sublethal concentrations of sodium arsenite on bluegills and the aquatic environment. Trans. Am. Fish. Soc., 95(3): 289–296.

Greenwood, P. H. 1958. The fishes of Uganda. The Uganda Society, Kampala: 124 pp.

Grizzell, R. A. 1967. Pond construction and economic considerations in catfish farming. Proc. Annu. Conf. Southeast. Assoc. Game Fish Comm., 21: 459–472.

Gruber, R. 1959. Toxicité pour *Tilapia melanopleura* de deux insecticides: l'Endrin et le Toxaphene. Bull. Agric. Congo Belge, 50(1): 131–139.

Guerrero, R. D. 1975. Use of androgens for the production of all-male *Tilapia aurea* (Steindachner). Trans. Am. Fish. Soc., 104: 342–348.

Halevy, A. 1979. Observations on polyculture of fish under standard farm pond conditions at the Fish and Aquaculture Research Station, Dor, Israel, during the years 1972–1977. Bamidgeh, 31(4): 96–104.

Henderson, C., Q. H. Pickering, and C. M. Tarzwell. 1959. Relative toxicity of ten chlorinated hydrocarbon insecticides to four species of fish. Trans. Am. Fish. Soc., 88(1): 23–32.

Henderson, C., Q. H. Pickering, and C. M. Tarzwell. 1960. The toxicity of organic phosphorous and chlorinated hydrocarbon insecticides to fish. *In* C. M. Tarzwell, Comp., Biological problems in water pollution 2. U.S. Dept. of Health, Education and Welfare, Robert A. Taft Sanit. Eng. Center, Cincinnati, Ohio: pp. 76–88.

Hepher, B. 1958. The effect of various fertilizers and their methods of application on the fixation of phosphorus added to fish ponds. Bamidgeh, 10(1): 4–18.

Hepher, B. 1959a. Use of aqueous ammonia in fertilizing fish ponds. Bamidgeh, 11(4): 71–80.

Hepher, B. 1959b. Chemical fluctuations of the water of fertilized and unfertilized fishponds in a sub-tropical climate. Bamidgeh, 11(1): 3–22.

Hepher, B. 1962a. Primary production in fishponds and its application to fertilization experiments. Limnol. Oceanogr., 7(2): 131–136.

Hepher, B. 1962b. Ten years of research in fishpond fertilization in Israel. 1. The effect of fertilization on fish yields. Bamidgeh, 14(2): 29–38.

Hepher, B. 1963. Ten years of research in fishpond fertilization in Israel. 2. Fertilizer dose and frequency of fertilization. Bamidgeh, 15(4): 78–92.

Hepher, B. 1966. Some aspects of the phosphorous cycle in fishponds. Verh. Int. Ver. Limnol., 16(3): 293–297.

Hepher, B. 1967. Some limiting factors affecting the dose of fertilizers added to fishponds, with special reference to the Near East. FAO Fish. Rep., 44(3): 1–6.

Hepher, B. 1975. Supplementary feeding in fish culture. Proc. Int. Congr. Nutr., 9(3): 183–198.

Hepher, B. 1978. Ecological aspects of warm water fishpond management. *In* S. D. Gerging, Ed., Ecology of freshwater fish production. Blackwell Sci. Publ., Oxford: pp. 447–468.

Hepher, B., and G. L. Schroeder. 1977. Wastewater utilization in Israel Aquaculture. *In* F. D'Itri, Ed., Wastewater renovation and reuse. Marcel Dekker, New York and Basel: pp. 529–560.

Hepher, B., J. Chervinski, and H. Tagari. 1971. Studies on carp nutrition. III. Experiments on the effect on fish yields of dietary protein source and concentration. Bamidgeh, 23(1): 11–37.

Hickling, C. F. 1960. The Malacca *Tilapia* hybrids. J. Genet., 57: 1–10.

Hickling, C. F. 1962. Fish culture. Faber and Faber, London. 317 pp.

Hickling, C. F. 1968. Fish hybridization. FAO Fish. Rep., 44(4): 1–11.

Hines, R. S. and D. T. Spira. 1974. Ichthyophthiriasis in the mirror carp *Cyprinus carpio* (L.) V. Acquired immunity. J. Fish Biol., 6(4): 373–378.

Hoffmann, W. E. 1934. Preliminary notes on the freshwater fish industry of South China, especially Kwantgunt Province. Lingnan University Science Bull., (5): 1–70.

Hora, S. L., and T. V. R. Pillay. 1962. Handbook on fish culture in the Indo-Pacific fisheries region. FAO Fish. Tech. Pap., (14): 204.

Hornell, J. 1935. Report on the fisheries of Palestine. Crown Agents for the Colonies, London: 106 pp.

Hruška, V. 1961. An attempt at a direct investigation of the influence of the carp stock on the bottom fauna of two ponds. Verh. Int. Ver. Limnol., 14: 732–736.

Huet, M. 1970. Textbook of fish culture, breeding and cultivation of fish. Fishing News Books, Surrey, England: 436 pp.

Hutchinson, G. E. 1957. A treatise on linmology. Vol. I. Geography, Physics and Chemistry. John Wiley and Sons, New York: xii + 1015 pp.

Ingram, W. M., and G. W. Prescott. 1954. Toxic fresh-water algae. Am. Midl. Naturalist, 52(1): 75–87.

Jalabert, B., P. Kammacher, and P. Lessent. 1971. Determinisme du sexe chez les hybrides entre *Tilapia macrochir* et *Tilapia nilotica*. Etude de la sexe-ratio dans les recroisements des hybrides de premiere génération par les espèces parentes. Ann. Biol. Anim. Bioch. Biophys., 11(1): 155–165.

Jhingran, V. G. 1974. A critical appraisal of the water pollution problem in India in relation to aquaculture. Proc. FAO Indo-Pac. Fish. Council, 15: 45–50.

Jhingran, V. G. 1975. Fish and fisheries of India. Hindustan Publ. Corp., Delhi: xi + 954 pp.

Johnson, D. W. 1968. Pesticides and fishes—A review of selected literature. Trans. Am. Fish. Soc., 97(4): 398–424.

Jones, R. O. 1965. Tolerance of fry of common warmwater fishes to some chemicals employed in fish culture. Proc. Conf. Southeast. Assoc. Game Fish Comm., 16: 436–445.

Kaneko, T. P. 1969. Composition of food for carp and trout. EIFAC Tech. Pap. 9: 161–168.

Kawamoto, N. Y. 1957. Production during intensive carp culture in Japan. Prog. Fish Cult., 19(1):26–31.

Kelly, H. D. 1957. Preliminary studies on *Tilapia mossambica* Peters relative to experimental pond culture. Proc. Conf. Southeast. Assoc. Game Fish Comm., 10:139–149.

Kerfoot, W. B., and G. A. Redmann. 1974. Permissible levels of heavy metals in secondary effluent for use in a combined sewage treatment-marine aquaculture system. I. Monitoring during pilot operation. *In* Wastewater in production of food and fiber—Proceedings. EPA-660/2-74-041: pp. 79–101.

Kerns, C. L., and E. W. Roelofs. 1977. Poultry wastes in the diet of Israeli carp. Bamidgeh, 29(4): 125–135.

Kessler, S. 1960. Eradication of blue green algae with copper sulphate. Bamidgeh, 12(1): 17–19.

Kesteven, G. L. 1942. Studies on the biology of Australian mullet. 1. Account of the fishery and preliminary statement of the biology of *Mugil dobula* Günther. Counc. Sci. Indus. Res. Aust. Bull., 157: 1–147.

Kimura, S., and Y. Matida. 1966. Study on the toxicity of agricultural control chemicals in relation to freshwater fisheries management No. 4. General summary of the studies on the toxicity of agricultural control chemicals for freshwater fishes by means of the bio-assay method. Bull. Freshw. Fish. Res. Lab. (Tokyo), 16(1): 1–10.

Kirk, R. G. 1972. A review of recent developments in *Tilapia* culture with special reference to fish farming in the heated effluents of power stations. Aquaculture, 1(1): 45–60.

Kuo, C.-M., C. E. Nash, and Z. H. Shehadeh. 1974. A procedural guide to induced spawning in grey mullet (*Mugil cephalus* L.). Aquaculture, 3(1): 1–14.

Lahav, M. 1971. Control of bugs in *Tilapia* spawning ponds (in Hebrew). Dag Hadio, (13): 13.

Lahav, M., and S. Sarig. 1969. Sensitivity of pond fish to Cotnion (Azinphosmethyl) and Parathion. Bamidgeh, 21(3): 67–74.

Lakota, S., A. Raszka, I. Kupczak, S. Hlond, J. Stefan, and J. Roszkowski. 1978. The effect of methoxychlor and propoxur on the health of carp fry (*Cyprinus carpio* L.). Acta Hydrobiol., 20(3): 197–205.

Leray, C. 1971. Experimental approaches to artificial feeding of some sea fish. *In* J. L. Gaudet, Ed., Report of the 1970 workshop on fish feed technology and nutrition. Resour. Bull., Bur. Sport Fish. Wildl., 102: 169–171.

Lellak, J. 1957. Der Einfluss der Fresstaetigkeit des Fischbestandes auf die Bodenfauna der Fischteiche. Z. Fisch.N.F., 6(8): 621–633.

Leventer, H., and J. Eren. 1969. Taste and odor in the reservoirs of the Israel National Water System. *In* Developments in water quality research. Proc. Jerusalem Int.

Conf. on Water Quality and Pollution Research, June 1969, Ann Arbor. Humphrey Sci. Pub., Ann Arbor: pp. 19–37.

Liao, I.-C. 1974. Experiments on induced breeding of the gray mullet in Taiwan from 1963 to 1973. Aquiculture, 2(2): 1–24.

Lin, S. Y. 1940. Fish culture in ponds in the New Territories of Hong Kong. J. Hong Kong Fish. Res Sta., 1(2): 161–193.

Lin, S. Y. 1949. Pond culture of warm water fishes. UN Sci. Conf. on Conserv. and Util. of Resour., UN Economic and Social Council, E/CONF.7/SEC/W.167 (mimeographed): 13 pp.

Ling, S. W. 1971 Travel report—Visit to Taiwan and Hong Kong. FAO, Rome. FID/71/R.7: 11pp.

Lovell, R. T. 1979. Flavour problems in fish culture. *In* T. V. R. Pillay and Wm. A. Dill, Eds., Advances in aquaculture. Fishing News Books, Farnham, Surrey, England: pp. 186–190.

Lowe (McConnel), R. H. 1955. The fecundity of tilapia species. East Afr. Agric. J., 21(1): 45–52.

Lowe (McConnel), R. H. 1958. Observations on the biology of *Tilapia nilotica* L. in East African waters (Pisces, Cichlidae). Rev. Zool. Bot. Afr., 57(1–2): 129–170.

Mabaye, A. B. E. 1971. Observation on the growth of *Tilapia mossambica* fed artificial diets. Fish. Res. Bull. (Zambia), 5: 379–396.

Mann, H. 1969. Geschmacksbeeinflussungen bei Fischen. Fette Seifen Anstrichm., Ernaehrungsindus., 71: 1021–1024.

Marek, M. 1975. Revision of supplementary feeding tables for pond fish. Bamidgeh, 27(3):57–64.

Marek, M., and S. Sarig. 1971. Preliminary observations of superintensive fish culture in Beith-Shean Valley in 1969–1970. Bamidgeh, 23(3): 93–99.

Marr, A., M. A. E. Mortimer, and I. Van der Lingen. 1966. Fish culture in Central East Africa. FAO, Rome: xiii + 159 pp.

Matida, Y., S. Kimura, and C. Yoshimuta. 1967. A toxic fresh-water algae, *Glenodinium gymnodinium* Penard, caused fish kills in artificially impounded Lake Sagami. Bull. Freshw. Fish. Res. Lab. (Tokyo), 17(2): 73–77.

Merla, G. 1966. Untersuchungen über die quantitative Entwicklung der natürlichen Nahrung der Karpfens (*Cyprinus carpio* L.) in Streck und Abwachsteichen und über ihre Beziehung zum Karpfenzuwachs. Z. Fisch., N.F., 14(3/4): 161–248.

Meschkat, A. 1967. The status of warmwater fish culture in Africa. FAO Fish. Rep., 44(2): 88–122.

Mironova, N. V. 1969. The biology of *Tilapia mossambica* Peters under natural and laboratory conditions (in Russian) Vop. Ikhtiol., 9(4): 628–639.

Mitchell, R. E., and A. M. Kirby. 1976. Performance characteristics of pond aeration devices. Proc. World Maricult. Soc., 7: 561–581.

Moav, R., and G. W. Wohlfarth. 1968. Genetic improvement of yield in Carp. FAO Fish. Rep., 44(4): 12–29.

Moav, R., G. Wohlfarth, and M. Lahman. 1960. An electric instrument for brand marking fish. Bamidgeh, 12(4): 92–95.

Moav, R., G. Hulata, and G. Wohlfarth. 1974. The breeding potential and growth curve differences between the European and Chinese races of the common carp. World congress on genetics applied to livestock production, 1: 573–578.

Moav, R., G. Hulata, and G. Wohlfarth. 1975. Genetic differences between the Chinese and European races of common carp. I. Analysis of genotype–environment interactions for growth rate. Heredity, 34(3): 323–340.

Moav, R., G. Wohlfarth, G. L. Schroeder, G. Hulata, and H. Barash. 1977. Intensive polyculture of fish in freshwater ponds. 1. Substitution of expensive feeds by liquid cow manure. Aquaculture, 10: 25–43.

Moriarty, D. J. W. and C. M. Moriarty. 1973. The assimilation of carbon from phytoplankton by herbivorous fishes: *Tilapia nilotica* and *Haplochromis nigripinnis*. J. Zool., Lond., 171: 41–55.

Mortimer, M. A. E. 1961. A handbook of practical fish culture for Northern Rhodesia. Dept. of Game and Fish., Lusaka: 150 pp.

Müller, W. 1959. Untersuchungen über Sonnenblumen-Extraktionsschrot als Karpfenfüttermittel im Jahre 1958. Dtsch. Fisch. Z., 6: 256–259.

Nakamura, K., M. Shimadata, H. Koyama, and H. Okubo. 1954. Fish production in seven farm ponds in Shioder Plain, Nagano Prefecture, with reference to natural limnological environment and artificial treatment. Bull. Freshw. Fish. Res. Lab. (Tokyo), 3(1): 27–29.

Nash, C. E., C.-M. Kuo, and S. C. McConnel. 1974. Operational procedures for rearing larvae of the grey mullet (*Mugil cephalus* L.). Aquaculture, 3(1): 15–24.

Neess, J. C. 1949. Development and status of pond fertilization in Central Europe. Trans. Am. Fish. Soc., 76: 335–358.

Nikolsky, G. V. 1963. The ecology of fishes. Academic Press, London and New York: 352 pp.

Nose, T., and S. Arai. 1979. Recent advances in studies of mineral nutrition of fish in Japan. *In* T. V. R. Pillay and Wm. A. Dill, Eds., Advances in aquaculture. Fishing News Books, Farnham, Surrey, England: pp. 584–590.

Odum, E. P., and A. A. De La Cruz. 1967. Particulate organic detritus in a Georgia salt marsh-estuarine ecosystem. *In* G. H. Lauft, Ed., Estuaries. AAAS Publ., 83: pp. 383–389.

Odum, H. T. 1956. Primary production in flowing water. Limnol. Oceanogr., 1(2): 102–117.

Odum, W. E. 1968. The ecological significance of fine particle selection by the striped mullet *Mugil cephalus*. Limnol. Oceanogr., 13(1): 92–98.

Odum, W. E. 1970. Utilization of the direct grazing and plant detritus food chains by the striped mullet *Mugil cephalus*. *In* J. H. Steele, Ed., Marine food chains. Oliver and Boyd, Edinburgh: pp. 222–240.

Otte, E. 1963. Die heutigen Ansichten über die Ätiologie der "Infektiösen Bauchwassersucht" der Karpfen. Wiener Tieraerztl. Wochenschr., 11: 995–10004.

Paperna, I. 1963. Dynamics of *Dactylogyrus vastator* Nybelin (Monogenea) populations on the gills of carp fry in fish ponds. Bamidgeh, 15(2/3): 31–50.

Parsons, D. A. 1949. Hydrology of a small area near Auburn, Alabama. USDA Soil Conservation Service and API Agric. Exp. Sta. SCS-TP-85: 40 pp.

Peleg, I. 1971. How to decrease the losses of carp fry from autumn spawnings (in Hebrew). Dag Hadio, (13): 9–12.

Perlmutter, A., L. Bogard, and J. Pruginin. 1957. Use of the estuarine and sea fish of the family *Mugilidae* (Gray mullets) for pond culture in Israel. Proc. Gen. Fish. Counc. Mediterr., 4: 289–304.

Pillay, T. V. R. 1973. The role of aquaculture in fishery development and management. J. Fish. Res. Board Can., 30(12), Pt. 2: 2202–2217.

Pillay, T. V. R. 1979. The state of aquaculture—1976. *In* T. V. R. Pillay and Wm. A. Dill, Eds., Advances in aquaculture. Fishing News Books, Farnham, Surrey, England: pp. 1–10.

Popper, D., and T. Lichatowich. 1975. Preliminary success in predator control of *Tilapia mossambica*. Aquaculture, 5(2): 213–214.

Porath, D., B. Hepher, and A. Koton. 1979. Duckweed as an aquatic crop. Evaluation of clones for aquaculture. Aquat. Bot., 7: 273–278.

Potts, W. I. M., M. A. Foster, P. P. Rady, and G. P. Howell. 1967. Sodium and water balance in the cichlid teleost *Tilapia mossambica*. J. Exp. Biol., 47(3): 461–470.

Pruginin, Y. 1956. Progress in methods of preventing wild fish from penetrating into fishponds. Bamidgeh, 8(3): 52–53.

Pruginin, Y. 1959. Equipment, construction and installations in fish culture. Proc. Gen. Fish. Counc. Mediterr., 5: 141–150.

Pruginin, Y. 1968a. Culture of carp and *Tilapia* hybrids in Uganda. FAO Fish. Rep., 44(4): 223–229.

Pruginin, Y. 1968b. Weed control in fish ponds in the Near East. FAO Fish. Rep., 44(5): 18–25.

Pruginin, Y., and A. Arad. 1977. Fish farming in Malawi. FAO, Rome. 22 pp.

Pruginin, Y., and A. Ben-Ari. 1959. Instructions for the construction and repair of fish ponds. Bamidgeh, 11(1): 25–28.

Pruginin, Y., and B. Cirlin. 1973. National status report on techniques used in controlled breeding and production of larvae and fry in Israel. FAO/EIFAC workshop on controlled reproduction of cultivated fishes, Hamburg, May 1973.

Pruginin, Y., S. Shilo, and D. Mires. 1975. Gray mullet: a component in polyculture in Israel. Aquaculture, 5: 291–298.

Rappaport, A., S. Sarig, and M. Marek. 1976. Results of tests of various aeration systems on the oxygen regime in the Ginosar experimental ponds and the growth of fish there in 1975. Bamidgeh, 28(3): 35–49.

Rappaport, U., S. Sarig, and Y. Bejarano. 1977. Observations on the use of organic fertilizers in intensive fish farming at the Ginosar station in 1976. Bamidgeh, 29(2): 57–70.

Reich, K. 1952. Some observations on the present state of pond fish culture in Israel. Proc. Gen. Fish. Counc. Mediterr., (2): 71–78.

Reich, K., and M. Aschner. 1947. Mass development and control of the phytoflagellate *Prymnesium parvum* in fishponds in Palestine. Palest. J. Bot. Jerusalem Ser. IV: 14–23.

Reinitz, G. L., L. E. Orme, C. A. Lemm, and F. N. Hitzel. 1978. Influence of varying lipid concentrations with two protein concentrations in diets for rainbow trout (*Salmo gairdneri*). Trans. Am. Fish. Soc., 107(5): 751–754.

Robson, T. O. 1974. The control of aquatic weeds—Mechanical control. *In* D. S. Mitchell, Ed., Aquatic vegetation and its use and control. UNESCO, Paris: pp. 72–84.

Rothbard, S., and Y. Pruginin. 1975. Induced spawning and artificial incubation of *Tilapia*. Aquaculture, 5(4): 315–321.

Safferman, S. R., A. A. Rosen, I. C. Mashni, and E. M. Morris. 1967. Earthly-smelling substance from a blue-green algae. Environ. Sci. Technol., 1: 429–430.

Sarig, S. 1962. Fisheries and fish culture in Israel in 1961. Bamidgeh, 14(4): 76–85.

Sarig, S. 1969. Fisheries and fish culture in Israel in 1967. Bamidgeh, 21(1): 3–18.

Sarig, S. 1970. Winter storage of Tilapia. FAO Fish. Bull., 2(2): 8.

Sarig, S. 1971. The prevention and treatment of diseases of warmwater fishes under subtropical conditions, with special emphasis on intensive fish farming. *In* S. F. Snieszko and H. R. Axelrod, Eds., Diseases of fishes, Book 3. T. F. H. Publications, Hong Kong: 127 pp.

Sarig, S. 1976. Fisheries and fish culture in Israel, 1975. Bamidgeh, 28(4): 67–79.

Sarig, S., and M. Marek. 1974. Results of intensive and semi-intensive fish breeding techniques in Israel in 1971–1973. Bamidgeh, 26(2): 28–48.

Schäperclaus, W. 1961. Lehrbuch der Teichwirtschaft. P. Parey, Berlin and Hamburg: v + 582 pp.

Schäperclaus, W. 1965. Etiology of infectious carp dropsy. Ann. N.Y. Acad. Sci., 126: 587–597.

Schroeder, G. L. 1974. Use of fluid cowshed manure in fishponds. Bamidgeh, 26(3): 84–96.

Schroeder, G. L. 1975. Some effects of stocking fish in waste treatment ponds. Water Res.., 9: 591–593.

Schroeder, G. L. 1975a. Nighttime material balance for oxygen in fishponds receiving organic wastes. Bamidgeh, 27(3): 65–74.

Schroeder, G. L. 1978. Autotrophic and heterotrophic production of microorganisms in intensely manured fishponds, and related fish yield. Aquaculture, 14: 303–325.

Shehadeh, Z. H. 1975. Report of the symposium on aquaculture in Africa, Accra, Ghana. FAO/CIFA Tech. Pap., (4): 36 pp.

Shehadeh, Z. H., C.-M. Kuo, and K. K. Milisen. 1973. Induced spawning of gray mullet *Mugil cephalus* L. with fractionated salmon pituitary extract. J. Fish. Biol., 5: 471–478.

Shell, F. W. 1968. Mono-sex culture of male *Tilapia nilotica* (Linnaeus) in ponds stocked at three rates. FAO Fish Rep., 44(4): 253–258.

Shelton, W. L., K. V. Hopkins, and G. L. Jensen. 1978. Use of hormones to produce monosex tilapia for aquaculture. *In* R. O. Smitherman, W. L. Shelton, and J. H. Grover, Eds., Culture of exotic fishes symposium proceedings. Fish culture section, American Fisheries Society, Auburn, Alabama: pp. 10–33.

Shiloh, S., and S. Viola. 1973. Experiments in the nutrition of carp growing in cages. Bamidgeh, 25(1): 17–31.

Sin, A. W. 1973a. The dietary protein requirements for growth of young carp (*Cyprinus carpio*). Hong Kong Fish. Bull., 3: 77–81.

Sin, A. W. 1973b. The utilization of dietary protein for growth of young carp (*Cyprinus carpio*) in relation to variations in fat intake. Hong Kong Fish. Bull., 3: 83–88.

Singh, V. P. 1980. Management of fishponds with acid sulfate soils. Asian Aquaculture, Philippines, 3(4): 4, 5–6.

Smith, E. V., and H. S. Swingle. 1941. The use of fertilizer for controlling several submerged plants in ponds. Trans. Am. Fish. Soc., 71: 94–101.

Soller, M., Y. Shchori, R. Moav, G. Wohlfarth, and M. Lahman. 1965. Carp growth in brackish water. Bamidgeh, 17(1): 16–23.

Spataru, P. 1976. Natural feeds of *Tilapia aurea* Steindachner in polyculture, with supplementary feed and intensive manuring. Bamidgeh, 28(3): 57–63.

Spataru, P., and B. Hepher. 1977. Common carp predating on tilapia fry in high density polyculture fishpond system. Bamidgeh, 29(1): 25–28.

Spataru, P., and M. Zorn. 1976. Some aspects of natural feed and feeding habits of *Tilapia galilaea* (Artedi) and *Tilapia aurea* Steindachner in Lake Kineret. Bamidgeh, 28(1–2): 12–17.

Stanhill, G. 1963. Evaporation in Israel. Bull. Res. Counc. Israel, II 6: 160.

Steffens, W. 1966. Trockenfuttermittel als Alleinfutter für Karpfen. Dtsch. Fisch. Ztg., 13: 281–289.

Strickland, J. D. H. 1960. Measuring the production of marine phytoplankton. Bull. Fish. Res. Board Can., 122: vii + 172 pp.

Suzuki, R. 1979. The culture of common carp in Japan. *In* T. V. R. Pillay and Wm. A. Dill, Eds., Advances in aquaculture. Fishing News Books, Farnham, Surrey, England: pp. 161–166.

Swingle, H. S. 1944. Construction of farm ponds. Agric. Exp. Sta., Alabama Polytech. Inst., Auburn, Alabama. Mimeographed Series, (8): 5 pp.

Swingle, H. S. 1961. Relationship of pH of pond waters to their suitability for fish culture. Proc. Pac. Sci. Congr., 9 (1957), 10 (Fisheries): 4 pp.

Szumiec, J. 1979. Some experiments on intensive farming of common carp in Poland. *In* T. V. R. Pillay and Wm. A. Dill, Eds. Advances in aquaculture. Fishing News Books, Farnham, Surrey, England: pp. 157–161.

Tal., S., and B. Hepher. 1967. Economic aspects of fish feeding in the Near East. FAO Fish. Rep., 44(3): 285–290.

Tamura, T. 1961. Carp cultivation in Japan. *In* G. Borgstrom, Ed., Fish as food. I. Academic Press, New York and London: pp. 103–120.

Tang, Y. A. 1970. Evaluation of balance between fishes and available fish foods in multispecies fish culture ponds in Taiwan. Trans. Am. Fish. Soc., 99(4): 708–718.

Tang, Y. A. 1979. Physical problems in fish farm construction. *In* T. V. R. Pillay and Wm. A. Dill, Eds., Advances in aquaculture. Fishing News Books, Farnham, Surrey, England: pp. 99–104.

Tapiador, D. D., H. F. Henderson, M. N. Delmendo, and H. Tsutsui. 1977. Freshwater fisheries and aquaculture in China. A report of the FAO fisheries (aquaculture) mission to China, 21 April–12 May, 1976. FAO Fish. Tech. Pap., 168: 83 pp.

Thaysen, A. C. 1936. The origin of an earthly or muddy taint in fish. I. The nature and isolation of the taint. Ann. Appl. Biol., 23: 99.

Thomson, J. M. 1963. Synopsis of biological data on the gray mullet *Mugil cephalus* Linnaeus 1758. CSIRO Fish. Oceanogr., Fish. Synop. 1. Melbourne, Australia: 72 pp.

Trewavas, E. 1965. *Tilapia aurea* (Steindachner) and the status of *Tilapia nilotica exul. T. monodi,* and *T. lemassoni* (Pisces, Chichlidae). Israel J. Zool., 14(1–4): 258–276.

Trewavas, E. 1968. The name and natural distribution of the *"Tilapia* from Zanzibar" (Pisces, Cichlidae). FAO Fish. Rep., 44(5): 246–254.

Trewavas, E. 1973. A new species of cichlid fishes of rivers Quanza and Bengo, Angola, with a list of known *Cichlidae* of these rivers and a note on *Pseudocrenilabrus natalensis* Fowler. Bull. Br. Mus. (Nat. Hist.) Zool., 25(1): 27–37.

Trewavas, E., J. Green, and S. A. Corbet. 1972. Ecological studies on crater lakes in West Cameroon. Fishes of Barombi Mbo. J. Zool. (London), 167: 41–95.

Truesdale, G. A., A. L. Downing and G. F. Lowden. 1955. The solubility of oxygen in pure water and sea water. J. Appl. Chem., Lond., 5: 53–63.

Uchida, R. N., and J. E. King. 1962. Tank culture of tilapia. U.S. Fish Wildl. Serv. Fish. Bull., 62: 21–52.

Vaas, K. F. 1948. Notes on freshwater fish culture in domestic sewage in the tropics. Landbouw (Batavia), 20: 331–348.

Vaas, K. F., and A. Vaas-van Oven. 1959. Studies on the utilization of natural food in Indonesian carp ponds. Hydrobiologica, 12(4): 104–109.

Van der Lingen, M. I. 1959. Some preliminary remarks on stocking rate and production of tilapia species at the Fisheries Research Centre. Proc. 1st Fish Day in S. Rhodesia, August 1957. Gov. Printer, Salisbury: pp. 54–62.

Van Someren, V. D., and P. J. Whitehead. 1959. The culture of *Tilapia nigra* (Günther) in ponds. II. Influence of water depth and turbidity on the growth of male *T. nigra.* E. Afr. Agric. J., 25(1–2): 66–72.

Viola, S., and G. Amidan. 1978. The effects of different dietary oil supplements on the composition of carp's body fat. Bamidgeh, 30(4): 104–109.

Viola, S., and U. Rappaport. 1977. Acidulated soapstocks in intensive carp diets—Their effect on growth and body composition. Intensive Fish Culture Station, Fisheries Dept., Ministry of Agric., Ginosar, Israel: 25 pp.

Walter, E. 1934. Gundlagen de allgemeinen fischereilichen Produktionslehre, einschliesslich ihrer Anwendung auf die Fütterung. Handb. Binnenfisch. Mitteleur., 4(5): 481–662.

Winberg, G. G. 1956. Rate of metabolism and food requirements of fishes. Nauchnye Trudy Belorusskovo Gosudarstvenoyo Universiteta imeni, V. I. Lenina, Minsk: 253 pp. Fish. Res. Board Can., Transl. Ser., 194 (1960).

Wohlharth, G. 1978. Utilization of manure in fish farming. *In* C. M. R. Pastakia, Ed., Proceedings—fish farming and wastes conferences. Inst. Fish Manage. and Soc. Chem. Ind., Univ. College, London: pp. 78–95.

Wohlfarth, G. W., and R. Moav. 1971. Genetic investigation and breeding methods of carp in Israel. Rep. FAO/UNDP/(TA), (2926): 160–185.

Wohlfarth, G. W., and G. L. Schroeder. 1979. Use of manure in fish farming–a review. Agric. Wastes, 1(4): 279–299.

Wohlfarth, G., M. Lahman, and R. Moav. 1962. Genetic improvement of carp. IV. Leather and line carp in fishponds of Israel. Bamidgeh, 15(1): 3–8.

Wohlfarth, G., R. Moav, and G. Hulata. 1975. Genetic differences between the Chinese and European races of the common carp. II. Multi-character variation—A response to the diverse methods of fish cultivation in Europe and China. Heredity, 34(3): 341–350.

Wolny, P. 1962. The use of purified town sewage for fish rearing (in Polish). Rocz. Nauk Roln. Ser.B., 81(2): 231–249.

Woynarovich, E. 1956. Die Organische Düngung von Fischteichen in produktions biologischer Beleuchtung. (English Summary). Acta Agronom. Acad. Sci. Hung., 6(3–4): 443–474.

Woynarovich, E. 1957. Bedeutung der Carbodüngung von Fischteichen. (The importance of carbon fertilization of fish ponds). Dtsch. Fisch. Ztg., 4(1): 15–16.

Woynarovich, E. 1962. Hatching of carp eggs in a "Zuger" glasses and breeding of carp larvae until an age of 10 days. Bamidgeh, 14(2): 38–46.

Woynarovich, E. 1975. Elementary guide to fish culture in Nepal. FAO, Rome: 131 pp.

Woynarovich, E. 1979. The feasibility of combining animal husbandry with fish farming, with special reference to duck and pig production. In T. V. R. Pillay and Wm. A. Dill, Eds., Advances in aquaculture. Fishing News Books, Farnham, Surrey, England: pp. 203–208.

Wunder, W. 1937. Richtlinien für die Verfütterung von Kartoffeln in Karpfenteich. Fisch. Ztg., 40(14): 3 pp.

Yaalon, D. H. 1964. Chemical changes in rain fed marsh water during the dry season. Limnol. Oceanogr., 9(2): 218–223.

Yashouv, A. 1969. Preliminary report on induced spawning of M. cephalus (L.) reared in captivity in freshwater ponds. Bamidgeh, 21: 19–24.

Yashouv, A. 1969a. The fishpond as an experimental model for study of interactions within and among fish populations. Verh. Int. Ver. Limnol., 17: 582–593.

Yashouv, A. 1969b. Mixed fish culture in ponds and the role of tilapia in it. Bamidgeh, 21(3): 75–92.

Yashouv, A. 1971. Interaction between the common carp (Cyprines carpio) and the silver carp (Hypophthalmichthys molitrix) in fish ponds. Bamidgeh, 23(3): 85–92.

Yashouv, A., and E. Berner-Samsonov. 1970. Contribution to the knowledge of eggs and early larval stages of mullets (Mugilidae) along Israel coast. Bamidgeh, 22(3): 72–89.

Yashouv, A., and A. Halevy. 1967. Studies on growth and productivity of Tilapia aurea and its hybrid "Gan Shmuel" in experimental ponds at Dor. Bamidgeh, 19(1): 16–22.

Yashouv, A., and A. Halevy. 1972. Experimental studies of polyculture in 1971. Bamidgeh, 24(2): 31–39.

Zismann, L., V. Berdugo, and B. Kimor. 1975. The food and feeding habits of early stages of gray mullet in Haifa Bay region. Acquaculture, 6(1): 59–75.

Index

Pages in *italics* indicate a more extensive discussion of the topic.

253